Let's Defeat Cancer!

THE BIOLOGICAL EFFECT
OF DEUTERIUM DEPLETION

GÁBOR SOMLYAI

Let's Defeat Cancer!
THE BIOLOGICAL EFFECT
OF DEUTERIUM DEPLETION

AKADÉMIAI KIADÓ, BUDAPEST

This book has been sponsored by
HYD Ltd. for Research and Development

First English edition

Published by Akadémiai Kiadó
P. O. Box 245. H-1519 Budapest
www.akkrt.hu

ISBN 963 05 7807 6

CONTENTS

TO THE READER
PREFACE TO THE ENGLISH EDITION

My book appeared in Hungarian only a year ago, but the publication dates of the Hungarian and English versions are separated by the turn of the century. We are just beginning to shape the twenty-first century and we sincerely hope that this English edition of the book will contribute to making the next hundred years less painful and agonising than the last. If there is a focus on the tremendous problem of cancer and allied diseases, I think we can reasonably hope for a breakthrough in the near future.

The book has summarised our research results and experiences of a decade; it contains all the necessary information to successfully use deuterium-depleted water in the course of cancer treatment. We have made great strides in Hungary during the past ten years: a growing number of veterinary surgeons daily employ our approved anti-cancer drug for animals (VETERA-DDW-25), and an increasing number of humans consume deuterium-depleted drinking water, marketed under the name Preventa-105. Unfortunately, the approval process for the human anti-cancer drug is still underway.

I will not go into details about the endless difficulties, failures and battles that have dogged my efforts to develop experimental proof, and then to make these new ideas known and accepted by the professional community; unfortunately without this acceptance and approval, my discovery cannot become useful for society.

I do hope that in Hungary the application of deuterium depletion in cancer therapy can proceed soon. I expect the English version of my book to help disseminate these ideas throughout the world, promoting the spread of the treatment and enabling hundreds of thousands of cancer patients to fight their disease with courage, confident of victory.

Budapest, 26 February, 2001

Gábor Somlyai

PREFACE

Cancer.

For most people it means a fearful disease; for the researcher it is a challenge; for the doctor it is part of everyday work; and for the tumour patient it is a question of life or death.

Although the disease is as old as mankind, it is only in the last decades that its significance has grown. In older times, when epidemics of cholera, smallpox, typhoid fever, and other diseases took their toll, relatively few people had died of tumorous diseases, a fact that can be attributed, primarily, to the shorter life span. In recent decades there are other diseases threatening the population: heart attack (infarction), cancer, and AIDS. Perhaps in the decades to come, heart attack, cancer and AIDS will be referred to just as cholera, smallpox and typhoid fever today, namely, as diseases to which science has found the key. At the same time it is sure that mankind later on will be confronted with numerous other diseases, which are still unknown to us.

It appears that while solving a number of problems, newer ones are being generated. With the development of technology tasks arise which only 50–100 years ago were unimaginable to be solved. When the train first appeared, passengers were asked to get off after a day's journey, as great speed was considered to be unhealthy; there was a time when someone had to walk in front of a car to call the attention of pedestrians to keep out of the way. Today we cannot help smiling at these stories. Trains rush at a speed of more than 200 km/h, cars flood highways and congest the streets of towns and to fly over continents takes but a few hours. Development is great but our fears have remained the same. In fact, it is only speed that has grown, the real essence of development, i.e. harmony with our environment, instead of growing has diminished – we are drifting ever further away from the desirable state.

During the last decades mankind has synthesised some 6 million different compounds of which only a few thousand are known exactly as to how they work and what their effects are. Strict rules have been set up to test a potential drug but all the laboratories of the world and several hundreds of years would not be sufficient to examine or exclude the consequences of possible counteractions of a single compound with all the other ones.

A requirement of harmonic development is solving a problem without generating newer ones. The results, observations, and advice included in this book serve as recommendations for the treatment of an old problem – cancer and tumorous diseases – but we do not know yet whether it generates new problems and if so, what these problems are. We can state, however, and with great certainty that, in line with the new principle, we do not want to poison the body with a synthetic compound alien to nature but to use the same method as nature does: to interfere with the deuterium-hydrogen metabolism of the cell.

By saying that this might be one solution of the cancer problem, we do not think that all patients will be cured from one day to the other, but the feasibility of a breakthrough is close.

When being asked about the significance of the discovery, I always say that it is something like trying to cure your fellow passenger's headache in a crashing aeroplane. The issue of cancer is but one among the numerous problems which is not worth much if we cannot solve the rest at the same time. Cancer, as a disease, in fact, is a signal, conveying the message that there are a lot of things on the planet Earth that we are performing very badly. Deuterium-depleted water (henceforth Dd-water) might treat the effect but it cannot solve the causes which are present in "developed" societies.

In this book we wish to speak about the facts; the experiments anybody can repeat; the joy and relief experienced by doctor and patient; and the fact that deuterium depletion may be an effective way of treating tumorous diseases.

With the help of Dd-water we would like to defeat this terrible disease, and not have to watch helplessly as our beloved one or friends fall victim to the disease.

Let us defeat cancer!

24 July, 1999

Gábor Somlyai

OUR MISSION

Our objective in publishing this book was to convey a certain attitude as well as offer our knowledge and experience that might help Hungary to lose its leading place in cancer mortality statistics. To reach this aim the co-operation of four groups of society is needed.

GROUP ONE

All those belong to this group who do not suffer from tumorous diseases. It is an obligation of the healthy population towards itself to decrease the probability of developing the disease. We try to help in this with advice and knowledge about the background of tumour growth. We will describe in detail how Dd-water – according to our present-day knowledge – can serve prevention.

GROUP TWO

Those professionals working in various fields of sciences, researchers (pathologists, biologists, microbiologists, geneticians, and molecular biologists) whose work is, in some way or another, linked to the research of tumorous diseases. They can help to discover the mechanisms which play an important role in the change of the deuterium concentration of cells, thus, in the regulation of cell division. So that they can be better informed we will provide the latest scientific results connected with deuterium depletion.

GROUP THREE

Doctors who fight a daily struggle with the disease. The application of Dd-water might open up new dimensions in tumour therapy. Their experiences and observations can convince them, within a short time, that together with the patient they can come out victorious in the war on cancer. To help them in their work the main rules of application of Dd-water will be outlined.

GROUP FOUR

To this group belong those who are in the most difficult situation – tumour patients themselves. They are desperate to have the present available therapies further extended. Unfortunately, in Hungary we suffer 33 thousand defeats every year in the war against cancer. An improved co-operation of doctors and patients to a great extent may contribute to the best possible outcome by using a combination of existing therapies and deuterium depletion. By sharing personal experiences the common wealth of knowledge may be enriched which, in turn, might lead to a more effective treatment of cancer.

The above groups could and should have done their best for the doctor and patient to come out victorious from the struggle against cancer. A sad experience of the last decades is that, for the time being, ours is a losing game. In spite of this, we think we can fulfil our mission. The basis for this was a simple scientific recognition. During the last 60 years – although it was known that the mass number 2 variant (isotope) of hydrogen, deuterium, occurs naturally – science has not attributed any significance to this fact in the regulation of biological processes.

The question is obviously simple: does naturally occurring deuterium have any role in the regulation of biological processes?

During the experiments the answer came within a short time: it definitely does.

This recognition enables a breakthrough in the field of treating tumorous diseases. We have not invented another, synthetically produced "miraculous molecule", we have just discovered a mechanism which is used by the cell itself to regulate division.

Book One

To say the truth is easy, but to make it plausible is difficult.

CHAPTER ONE

To Everybody

INTRODUCTION

In our days when children use the computer with great skill and routine, when we can program our mobile phones and videos and surf the net, it is most distressing that we are not aware of the functioning of our own bodies.

We do not need a detailed knowledge of biology, we only have to pay attention to the causes and correlations which may have a key role in decreasing the probability of tumour development. In this chapter we wish to outline the development and treatment of cancer as well as the possibilities provided by deuterium depletion for those who have no deep knowledge of the results of biology, medicine or molecular biology. We do not intend to convey knowledge difficult to understand, but we definitely aim to answer some simple questions and to make it clear that it is possible to improve our chances of avoiding and defeating tumorous diseases.

ON TUMOURS IN GENERAL

WHAT IS CANCER?

It is difficult to define in a short but professionally exact way the disease called "cancer" in everyday terms. Strictly speaking, the expression "cancer" (carcinoma) refers to malignant tumours originating from epidermic cells only, but commonly this word refers to all malignant tumorous diseases. A common characteristic of malignant tumorous diseases is that they originate from a single cell, and after a certain time (which can be several weeks but mostly 3 to 5 years) a group of cells comes into being which significantly differs from its environment both in function and structure. Its most typical feature is uncontrolled division, resulting in these cells to branch out from the basic tissue and grow into surrounding tissues. Cells detached from the tumour get into the blood or lymphatic stream and thus to other parts of the body where they settle and proliferate creating a new tumour (metastasis) or, in the case of the disease of hematopoietic organs, flood the bloodstream with a mass of immature (blastoid) cells.

On the average, it takes 4 to 5 years for a single cell to grow into a tumour of one centimetre, detectable with present-day diagnostic devices. At this stage the number of cells building the tumour is above 10 million. The insidiousness of cancer is, in the first place, due to this fact. In the first 4 to 5 years, when the size of the tumour cannot be detected and the patient may be totally free of symptoms and complaints, numerous cells can break loose from the tumour. If one of them successfully settles in some place in the body, in a further 4 or 5 years the patient may develop another tumour even if the primary one was removed years earlier. (This is why we consider the five year symptom-free period of tumour patients decisive, because if no tumour appears within this period, it can be supposed with great probability that with surgical and other treatments all tumorous cells in the body were successfully eliminated.)

On the basis of these facts, the disease might seem predestined or inevitable but luckily this is not quite so. A tumorous cell must fight for its survival just like any other living being in nature. The defense system of the body strictly controls the continuously developing malignant cells. Health is a dynamic state of balance. Disease appears when this balance is upset and becomes unstable for a longer period. It is most important to strive in the course of our lives to prevent our physical and psychic balance from becoming upset. If, however, the tumour has already developed, we should try to do everything in order to restore the balance in our bodily processes.

WHY DO TUMOURS GROW?

The tumour is the "end-product" of a very complex process consisting of multiple steps. It can be likened to a "ball" symbolising the cell in its normal state (healthy cell), which has rested in a small recess of a hillside. Processes occurring within the cell sometimes result in the centre of gravity of the cell to almost topple over that small brim which prevents it from "rolling down" the slope but further processes might help it get back to a safer point of the recess. In healthy cells, forces aligned in the direction of the slope are in balance with those drawing it back. The first event leading to tumour development is when the cell topples over the first brim. To prevent the cell from breaking away there is another recess in which it again spends some time but if the forces aligned in the direction of the slope prevail, it soon topples over again and starts rolling downhill. The more a cell overcomes such obstacles the lower the brim on the next level and the slope will also get steeper. If the processes occurring within the cell cannot halt it at any of these points permanently, the cell breaks away and it cannot be stopped – it starts to divide beyond control *(Figure I.1)*.

In the following we present some processes which, when they are imbalanced, result in the cell starting to roll "downhill".

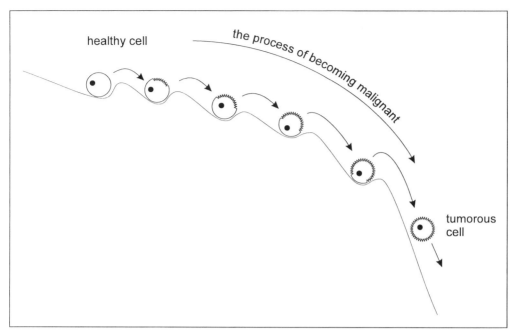

Figure I.1 *The growth of a tumour is a process of several steps. The picture of a "slope" demonstrates how a cell becomes more malignant in the course of several years, until, losing control, it breaks away and rolls down the slope where nothing is in the way of its division.*

DEFECTS IN THE GENETIC PROGRAM

It is generally accepted that the primary cause of tumour growth is a series of accidents in the genetic program. During the life span of humans, from insemination to death, about 10^{16} cell divisions occur. Each cell (almost each, to be exact), contains a genetic program consisting of 3 billion "letters". The program is written by Life with the help of 4 "letters" only (A, T, G, C) which correspond to the four bases that constitute DNA, the hereditary substance of cells. (This book contains approximately 260,000 letters, so the genetic program of a single cell could be written down in more than 10,000 books of the same size.) Before each division, the cell doubles its genetic program so that the successor cell should also possess the same program. When replicating the genetic program, the enzyme (protein) which carries it out makes errors with a certain frequency, so in the successor cell it is not always the appropriate "letter" that gets to a given place of the genetic program. **If these accidents occur at certain points of the genetic program consisting of 3 billion "letters" (base pairs), i.e. in sections which have a distinguished role in regulating cell division, then these cells behave differently from the surrounding ones and divide more often.** This can result in the cell group outgrowing its surroundings, the visible appearance of which is the tumour itself.

19

Life could not have reached the present degree of complexity had the so-called repair mechanisms not developed whose "task" is to immediately remedy the errors in the course of replicating. Thus, the "net" balance of actual genetic defects can be defined as the resultant of the number of occurring errors and the efficacy of the repair system. This works so well that while during the copying of DNA one error occurs in about every thousand "letters", thanks to the repair mechanism the frequency of defects in the whole of the DNA copy is less than one from among one million bases which means that out of a thousand defects only a single one cannot be corrected by the cell. The probability of developing a tumorous cell depends on the frequency of errors in the genetic program as well as on how precisely and in what percentage the repair mechanism corrects them. *Figure I.2* demonstrates the factors affecting the origin of genetic errors and those eliminating them. (It is important to know that just as people differ in many respects, the same is true for the activity of the "repair" enzyme. There are people with whom this mechanism works with a high efficacy and others whose percentage of defects is higher. This, in

Figure I.2 *From the point of view of tumour growth, the primary process is the defect in the genetic program. This is a natural concomitant of life, in the course of all cell divisions, from among one million base pairs one genetically defected gets into the successor cell (A). During our lifetime the frequency of defects may significantly increase if carcinogens find their way to the cell (B). In some cases the abundance of genetic accidents may result in the malfunction of the repair system which considerably increases the probability of tumour growth (C).*

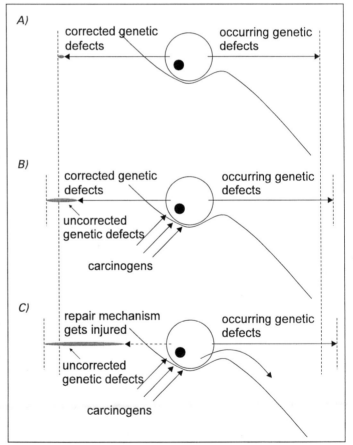

part, explains why, from two persons belonging to the same risk factor group, one becomes a cancer patient and the other does not.)

Part A) of *Figure I.2* shows that in a person belonging to a group of low risk factor, the length of the two arrows is about the same, although the one pointing in the direction of repair is somewhat shorter, which is explained by the fact that even the most exact repair mechanism is unable to correct one out of a thousand genetic defects (occurring on a sequence of a length of approximately 1 million base-pairs). As a result, during our lifetime, the number of genetic defects, if slowly, but increases. Luckily, however, the occurrence of a single genetic defect at a certain place does not necessarily and immediately result in a tumorous cell. (The cell cannot, at once, "roll down" the slope because the "recesses" make it stop.) This is the consequence of the fact that although a single genetic defect can trigger the cell to become malignant, for this to happen, further genetic accidents must occur in the cell (it must topple over further brims).

In part B) of *Figure I.2* we try to demonstrate how we ourselves tip over this delicate balance of Nature and push it in the direction of the "slope". On the impact of certain materials, genetic defects may significantly increase in number. This explains why among persons working with carcinogenic materials, smokers, or those exposed to a strong UV radiation, certain tumorous diseases occur more frequently. Obviously, with the increase in the number of genetic defects the chances of the repair mechanism to "go wrong" become higher. Part C) of *Figure I.2* helps to illustrate what happens to the cell that has lost its mechanism to prevent it from "rolling down". These cells cannot be halted, and within a short time they lose all control.

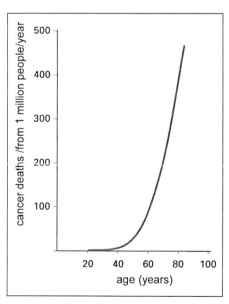

Figure I.3 *Cancer mortality rate as a function of age. It is clearly visible that from the age of forty the death rate increases dramatically. From among the 50-year-old population 30, while in the case of the 80-year-old population 400 cancer deaths occur in one million. The graph also shows why tumorous diseases were of less significance at the beginning of the 20th century, when life expectancy was merely 40–45 years. The development of civilisation and life expectancy extended to 70–80 years has resulted in a rapid increase in the number of cancer incidences.*

21

From the point of view of prevention, these observations convey two essential messages *(Figure I.3)*.

a) **All alien materials getting into the body which increase the incidence of genetic defects (mutation) increase the probability of cancer.** This explains why, in the case of smokers, those living in polluted areas or working with carcinogenic materials, the incidence of tumorous diseases differs from that of the average population.

b) **The younger the age when the first genetic defects occur, the greater the probability of "collecting" further mutations which lead to tumour growth in the course of a lifetime.** This is supported by the fact that the probability of cancer incidences increases with age. [*Source:* Molecular Biology of the Cell, 2nd edition.]

In the course of our lifetime, ever more cell divisions take place and the number of randomly occurring genetic defects in the genetic program also increases. The result of this abundance is that within a cell, where the genetic accident affecting the regulation of cell division had formerly happened, the appearance of a further one can easily tip the cell over the last brink.

Being aware of the above, it is especially sad to see children and teenagers smoke their first cigarettes early in the morning on the way to school. If a more conscious and vigorous health campaign will not change this, it is highly probable that many of them will suffer from cancer and that the disease will develop at an early age.

The risk of persons whose cells have a given genetic defect at the moment of birth is similarly greater. Today, the susceptibility in many families to develop cancer can be well traced, so it is especially important that they reduce the likelihood of further genetic defects occurring.

Cells not only try to correct genetic defects but there are also refined mechanisms to decrease the amount of reactive materials within the cell, which may cause genetic damage. It is obvious that by keeping the intactness of DNA, what a great advantage is attained in selection if the repair mechanism has to correct lesser errors only. The cells, in which these mechanisms are present, have an undisputable advantage over the rest. During the natural processes of cells, great masses of reactive material come into being every day as well as a number of harmful materials from the environment, which find their way into the cells. Well-known cell protectors are the anti-oxidants (Vitamins C, E, and A as well as selenium) whose role, among others, is to neutralise "free radicals" emerging in the cells. Cells also produce numerous proteins to protect them from materials of other types. Such is, for instance, the enzyme called N-acetyl-transferase (NAT) or the enzyme glutathione-S-transferase M1 (GSTM1) which in case of a smoker may have a decisive role in tumour growth. Smokers with low blood levels of NAT have a 2.5 times greater risk of developing cancer of the bladder than smokers with a high NAT activity. Similarly, insufficient levels of GSTM1 trebles the risk of lung cancer. If we imagine the cell on the slope, we can see that this state is also the result of a balancing process. If a person has low levels of the above enzymes, the balance will not necessarily tip over if he quits smoking, i.e. it does not burden his weak protective system any further *(Figure I.4)*. The reverse is also true: not all smokers develop lung cancer, a fact that is partly due to the strong detoxifying system which protects the genetic program from getting damaged *(Figure I.4.B)*. Unfortunately it is also true that the cell cannot remain on the slope of the hill if its weak detoxifying system is bombarded by a mass of reactive molecules *(Figure I.4.C)*.

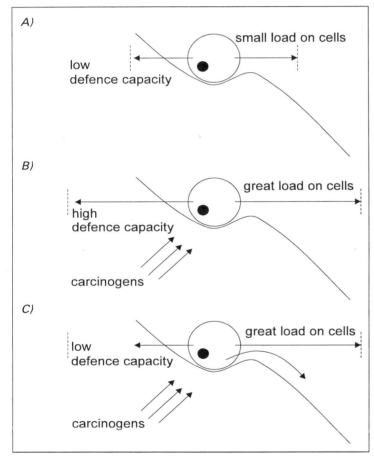

Figure I.4 *In every second, millions of chemical reactions take place in the cells. In this process a great quantity of compounds and reactive materials come into being which may damage the cell. During evolution certain mechanisms and enzyme systems appeared which had to neutralise materials dangerous for the cell. In an ideal case this capacity of the cell is sufficient for neutralising the decisive part of reactive materials if the load on cells is small (A) or if the load is great but the "detoxifying system" is strong enough (B). The balance tips over if a great load is put on the cell along with a weak defence system (C).*

THE PROCESS OF THE CELL'S TURNING MALIGNANT

As mentioned before, in the course of DNA replication, genetic accidents occur with certain frequency and their number further increases if the cell is exposed to harmful materials and/or if the detoxifying system of the cell is weak. From the point of view of cancer growth, genetic defects occurring in the so-called oncogenes, i.e. DNA sequences, which have a key role in the regulation of cell division, are especially significant. Researchers have found, however, that the appearance of oncogenes does not necessarily mean the appearance of a tumour. It has also become clear that not only oncogenes exist but genes as well, whose activity suppresses the activity of oncogenes. These are the tumour suppressor genes.

In fact, tumour growth can be traced back to disturbances in the information processing of the cell. Normally, the cell divides when it gets an outside signal to do so, with the help of the binding of a growth hormone. Encouraged by oncoproteins (proteins "produced" by the cell on the basis of oncogenetic codes coming from a defective genetic program) tumour cells may lose their dependence from growth factors. One way for this to happen is that the cell itself starts to produce the growth factor flooding the surrounding cells with them, in this way stimulating its own division. Another possibility is when the so-called "receptor", which binds growth factors, changes as a result of a genetic accident. From then on, the receptor starts sending messages to the inside of the cell as if it received an order to divide although no binding of the growth hormone occurred.

The functioning of suppressor proteins is not as clear yet as that of oncoproteins but it can be stated, in general, that these proteins suppress signals stimulating growth.

The above are, naturally, not the only two proteins active in cells, there is also a complex system of numerous other proteins. Their task is to perceive, convey and process the mass of signals arriving every moment from the environment, to consider the force and direction of the signals and, accordingly, to trigger or suppress cell

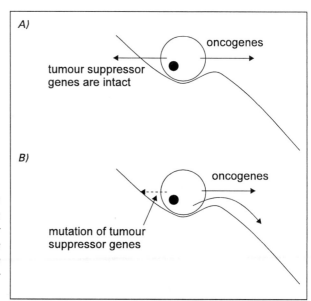

Figure I.5 *The effect of oncogenes arising through mutations is counterbalanced by tumour suppressor genes (A). The balance tips over when, by mutation, the cell loses its tumour suppressor genes, too (B).*

division. Proteins co-operate with each other and "harmonise decisions". It is also obvious that we can speak of co-operation and, through it, the maintenance of balance in this respect only until the tumour suppressor protein is present *(Figure I.5. A)*. A decisive point in the appearance and growth of a tumour is when the cell loses one or more of its tumour suppressor proteins *(Figure I.5.B)*.

APOPTOSIS WILL NOT START

It is easy to see that the net proliferation of a given cell group is composed of two factors: how frequently do new cells come into being and how frequently do old ones become extinct. In healthy tissues this process is balanced throughout the whole life of a person. This is made possible because a cell not only has a very exact regulating system to trigger cell division but also one to regulate when a cell should "die". Scientists call this process apoptosis or programmed cell death, because the program "drives the cells to commit suicide". We have mentioned before that in the course of DNA replication, errors may occur in the genetic program with a certain frequency. Our body protects itself from cells carrying multiple genetic defects so that after a certain number of divisions the cell destroys itself thus freeing the body from possible later tumour cells. Unbelievable as it may sound, the cells possess a counter which signals how many times the cell has divided and if this number reaches the critical value the suicidal program starts: within a short time the cell falls apart and disappears. The cell in some way or another also perceives if DNA has been damaged seriously. In such cases it does not even try to repair the defects but destroys itself. The above mechanism works as a twofold filter against the appearance of a tumour. Firstly, it prevents genetically impaired cells from surviving but also, if the primary tumour has already appeared, actual cell proliferation only occurs if cells succeed in eliminating the program triggering their death. Otherwise cells divide in vain as, because of the high rate of cell death, there will be no net cell number increase or, if so, a minimal one. Based on the above, I think now it is obvious to everybody that it is only a matter of time that, if from among the cells in the primary tumour at least one can eliminate the apoptosis program, this means such an advantage in selection that it enables the cell to quickly outgrow its surroundings and the primary tumour starts to grow at a great speed *(Figure I.6, Phase III)*.

NUTRIENTS AND OXYGEN SUPPLY OF THE PRIMARY TUMOUR

The above chain of events happening at a molecular level is a process that can last for years. At this stage the size of the primary tumour has not yet reached 1 millimetre and consists of maximum some 10 thousand cells. The genetically defective cells, which could rapidly divide, are given but having reached this size the cell group has to fight still another battle. The nutrients and oxygen supply of cells constituting the primary tumour can no longer be solved by diffusion as until now, so the growth of the cell group comes to a halt again. Its further growth is made possible only if the blood supply of the cell group is ensured. The lack of it, in some cases, can further delay the formation of a palpable tumour for years. One or more cell can, with time, however, "direct" the growth of capillaries from surrounding tissue into the primary tumour. It is at this phase that the last obstacle in the way of tumor growth is overcome, the nutrients and oxygen supply of the tumour is cared for and the number of tumour cells begins to multiply explosively *(Figure I.6, Phase IV)*.

We stress again that in healthy persons the counteracting processes are in balance at various levels *(Figure I.6)*. Genetic defects do occur but their majority are repaired; oncogenes come into being but their activity is compensated for by tumour suppressor genes; the increase in the number of genetic defects could lead to the appearance of tumour cells but the cell starts its own suicidal program thus freeing the body from genetically impaired cells. If the system is efficient enough it manages to keep the number of malignant cells continuously low. This is confirmed by the fact that examining the prostates of elderly deceased men, researchers have found that – in compliance with their age – in a significant part of them (60–70 percent of the 60 to 70 age group) the presence of tumorous cells could be detected. However, only a very low percent died in prostate cancer because in most cases the process was under the control of the human organism.

Phase I – Defects in the genetic program

In the course of cell division, in every 1000 base pairs approximately one genetic accident occurs and from among one thousand of these 999 are corrected by the repair mechanism. Thus in one million base pairs there finally is but one, which is genetically defective. This means that in the human genetic program of 3 billion base pairs, with each cell division there occur approximately 3,000 genetic defects. The actual number of genetic defects is increased by the intracellular accumulation of reactive material, which may generate genetic defects. The number decreases if the strong detoxifying system of the cell neutralises them and the highly effective repair mechanism of the genetic program corrects the defects.

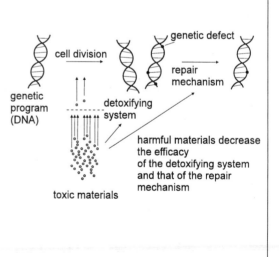

Figure I.6

cell nucleus

defective cell
in the part of the
genetic program
responsible
for division

a more frequent division
than in healthy cells

in the course of division
newer defects occur,
thus the ability to divide increases

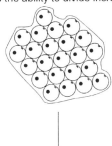

healthy cells turn into malignant ones
dividing beyond control

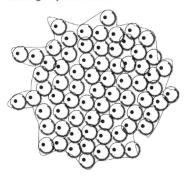

the frequency of apoptosis
defines the net growth
of the cell group

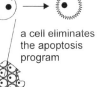

a cell eliminates
the apoptosis
program

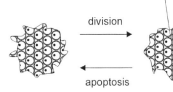

division

apoptosis

As a resultant of the above processes within a certain time if the detoxifying system is weak, the harmful materials are considerable and the repair mechanism is functioning at low efficacy then a genetically defective cell may appear sooner, if, however, the carcinogen level is low and the repair mechanism works properly then later. This genetic defect had occurred in a section which had a special role in cell division in consequence, the cell began to divide more frequently than the surrounding healthy ones.

Phase II – The process of the cell turning malignant

As a result of the "events" in Phase I, within a given organ a cell containing a genetic defect might appear, and due to this, the control of cell division is lost and the number of cells carrying this genetic defect begins to increase. In the course of further divisions, from the point of view of division regulation, newer genetic defects appear (mutations), which result in a further modification of the cell characteristics: its ability to divide increases. Due to these processes the formerly healthy cell turns into an uncontrollably dividing malignant one. It is especially important when, and in what oncogene or tumour suppressor gene, the defect occurs.

Phase III – Apoptosis will not start

The net growth of a given cell group is defined by how frequently the cells divide and how often they become extinct. Apoptosis is one of the emergency possibilities for the body to get rid of its genetically defective cells which have become malignant.

According to the law of averages, if primary tumours occur often enough and the number of cells in them reaches a given value, the likelihood of a cell to occur which inhibits the apoptosis program will increase and the growth of the primary tumour will no longer be hindered by the fact that some parts of it have died. It is then that the cell group can start to grow really quickly.

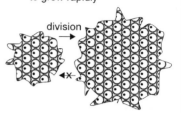

the primary tumour starts to grow rapidly

division

Phase IV – Nutrient and oxygen supply of the primary tumour

Quick growth lasts only until the tumour reaches the size of 1 millimetre i.e., until the cells within the primary tumour can get nutrients and oxygen by diffusion. At this stage growth stops and the tumour might not develop further if blood supply is not secured. One or more of the cells in the primary tumour may, in time, be able to "direct" the growth of capillaries from surrounding tissues to the primary tumour. Thus, the nutrient and oxygen supply of the tumour is secured and the number of tumour cells begins to increase at a great speed.

the nutrient and oxygen supply of the primary tumour is made possible through diffusion only until it is less than 1 millimetre

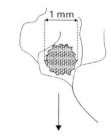

1 mm

some cells provide the nutrient and oxygen supply

Phase V – The growth of the tumour

The growth of the cell group may continue for years without the appearance of any symptoms. The majority of tumours are diagnosed and controlled medically only when their size is about one centimetre. At this stage the tumor already consists of over tens of millions of tumour cells.

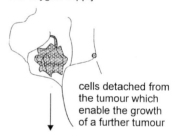

cells detached from the tumour which enable the growth of a further tumour

Phase VI – Metastases

Unfortunately, it is a common phenomenon that although the patient has undergone a successful surgery, after some months or years a new tumour appears, occasionally at a totally different point of the body than the original tumour had been. With some exceptions, however, this is not a new tumour but a metastasis of the one formerly removed. As cells detach from the primary tumour, through the blood or lymphatic stream they reach distant parts of the body and create a new tumour. If they manage to survive after their "wandering" and to successfully settle, in some cases it is only 4–5 years later that a new tumour appears, a metastasis of the primary one.

the number of tumour cells begins to grow dramatically

1 cm

beginning of medical intervention

To prevent the appearance of tumorous diseases, our aim should be to maintain the balance of our organism throughout our lives. This is rather difficult to ensure in our days when the amount of carcinogens in the environment continuously increases; our life is under constant stress; and the lack of exercise and inappropriate nutrition and a number of other factors also favour the development of tumours. Statistically, in our "developed society", this is manifested by the increasing number of cancer patients and growing mortality. This is also supported by the statistical data indicating that during the past 20 years, among 40 to 49 year old males the occurrence of cancer grew 200-fold.

The tumour appears in the body as an "end-product" of a process lasting for many years. Each phase of its development is characterised by the quantitative accumulation of genetic defects, some of which assume special significance: cells of this kind acquire a unique potentiality as opposed to other cells. In fact, the development of a tumour rarely occurs and can be led back to a chain of successive, random events. In the following we wish to demonstrate the causes and the phases of its development *(Figure I.6)*.

WHY COULD OUR GRANDFATHERS SMOKE?

People (mostly smokers) often say that smoking is not responsible for tumour growth and they keep referring to their 80 or 90-year-old grandfathers. Unfortunately, this argument is no longer valid. In the case of our forefathers the only risk factor was, in fact, smoking, as in those times people lived in a healthy environment and were not subject to the stress of big cities and to polluted air, so their organisms were able to keep a balance despite the existence of tumour cells. In our days, when the bulk of the population lives in big cities and in addition in this polluted environment most people smoke, harmful effects multiply and this, in too many cases, is beyond the capacity of the defence system. Consequently, the tumour gets out of control as we have tried to demonstrate in previous figures. The increase of risk factors is well indicated by the fact that among those who work with the very dangerous material, asbestos, the risk of lung cancer is 16 times higher and if, in addition, the endangered person smokes, this number is 600 times more.

WHICH WAY IS CANCER THERAPY GOING IN OUR DAYS?

The present state of the war against cancer can be evaluated from various aspects. One favourable approach is that tumorous diseases should not automatically be looked upon as lethal ones. In some respects, medicine has, for the last decades, been developing at an immense pace and in the case of certain tumours (tumour of the testicles or childhood leukemia) following appropriate treatment, a significant percentage of patients can be considered cured. Another achievement is when, though not cured, the patient's life can considerably be prolonged (unfortunately often along with a considerable deterioration in the quality of life).

In terms of numbers: it is a fact that since 1993, in Hungary alone, 200,000 cancer patients have been buried. As if, during 7 years, four smaller towns had become extinct.

The question arises how such statistical data can verify the achievements of cancer therapy. **If we are unable to face the facts and accept that all endeavours up till now and the further development of current methods of treatment applied so far will not lead us to solve the problem, then we will not be able to change the situation and find realistic answers to the problem**. The consequence of being too ready to accept results as positive is that we turn away from possibilities that offer themselves instead of reaching out toward them, although we are badly in need of every idea, every new concept. This, naturally, does not mean that every new thought carries in itself a real breakthrough, but chance must be given to every new idea and, thereby to our survival.

Before I could be accused of making exaggerated statements regarding the present situation, let me quote some paragraphs by Tim Beardsley, writer on the editorial staff of *Scientific American*. The title of the article published in the January 1994 issue of the journal reflects the actual situation: "A War Not Won".

"There was so much good news at their meeting in September that the members of the President's Cancer Panel might have been pardoned had they been overwhelmed by euphoria. Reports of promising therapies and diagnostic manoeuvers swirled with accounts of deep new insights into the underlying genetics and molecular biology of the disease. Then the mood grew somber – or was it tense? John C. Bailar, a noted professor of epidemiology and biostatistics at McGill University, began his observations on recent trends in the morbidity and mortality of cancer.

Bailar had created a storm in 1986 after publishing a damning lack-of-progress report on the 'war on cancer' initiated by the President Richard M. Nixon when the chief executive signed the National Cancer Act in 1971. Bailar's unflinching summary of the latest body counts last fall, part of a formal evaluation of the national cancer program, led to the same disturbing result. 'In the end, any claim of major success against cancer must be reconciled with this figure,' he said, pointing to a simple graph that showed a stark continuing increase in U.S. death rates from cancer between 1950 and 1990. 'I do not think such reconciliation is possible and **again conclude, as I did seven years ago, that our decades of war against cancer have been a qualified failure**'[...]

The numbers based on data supplied by the National Cancer Institute (NCI), indeed present a grim picture. Bailar's principal conclusion, with which the

NCI agrees, is that U.S. cancer death rates went up by 7 percent between 1975 and 1990. This number, like all those Bailar cited, has been adjusted to compensate for the changing size and composition of the population with respect to age, so the increase cannot be blamed on people's dying less often from other diseases. Cancer is the second leading cause of death in the U.S. after heart disease; 526,000 cancer deaths were expected to have occurred in the U.S. in 1993.

What do these facts mean? Would analysis of them reveal environmental or other influences that are triggering the disease? Do the results justify the massive, frontal assault, which relies on research into the fundamental causes, on efforts to devise sensitive diagnostic procedures and on empirical attempts to develop cures? **Have the researchers and clinicians been barking up the wrong trees for the past two decades?**[...]

Thus, the war against cancer is one in which the foe's order of battle changes constantly. **Death rates have declined for such cancers as those of the colon and rectum, stomach, uterus (including the cervix), bladder, cranium, bone, gallbladder and testis. Death rates from cancer in children fell by almost half between 1973 and 1989,** in large part because of better treatments. Yet because cancer in children was rare to begin with, this improvement – and smaller gains in young adults – has had only a minuscule effect on the big picture. **Overall, the increases in cancer death rates are twice as large as the decreases**[...]

In principle, death rates should reflect both improvements in treatment and changes in incidence. Unfortunately, colorectal cancer is only one of the big four killers – lung, colorectal, breast and prostate – that is becoming more curable. By a cruel twist of fate, the other cancers that can now be cured somewhat more successfully than in 1973 are relatively rare. They are Hodgkin's disease, some leukemias, cancer of the thyroid, testicular cancer and, perhaps, cancers of the uterus and bladder[...]

A completely different way to look at progress in the war against cancer is to examine what proportion of people diagnosed with different types survive for at least five years. Although five-year survival rates say nothing about incidence, they are seen as highly relevant by the people diagnosed as having cancer. Unfortunately, the comfort that these data offer is distinctly cold. **Despite the heartening gains in cure rates in the young, the NCI estimates the overall improvement in five-year survival since Nixon launched the program is only 4 percent.** Bailar told the President's Cancer Panel that even that figure may constitute an overestimate[...]

Optimists, such as Greenwald, feel that large-scale improvements in death rates and incidence rates will eventually result from the years of investment in basic cancer research, although Greenwald guesses it may be a decade before the benefits are widely felt. Gene therapy, immunetherapy and antisense RNA technology are just a few of the methods now in early testing. Such subcellular interventions could, according to their proponents, bring about dramatic gains by changing the activity of specific genes in tumors and selectively modifying the immune response. But skeptics, including Bailar, say they have heard that kind of talk too many times before. **'No knowledgeable person can continue to believe there is necessarily a spectrum of marvelous cures of cancer waiting to be found'**, he asserts. Bailar says he is fed up with the 'constant procession of hopeful new stories' suggesting a cure is just around the corner.

Existing chemotherapies, despite improvements, are still arduous. Oncologists who employ them know they are wielding double-edged swords. Some of the treatments for lymphoma and leukemia trigger other cancers, after therapy for the primary disease has been successfully completed[...]

'If there has been a change it is that **we must shed our illusions that cancer is an easy problem. It is a formidable problem.'**

Bailar is among those who believe the NCI should devote much more of its money to prevention. **'We've come to the point where we must face a really serious problem square in the face. What if there aren't any major advances to be obtained in chemotherapy?'** he demands. **'For a lot of years now, we've been tinkering. It's not going to solve the big problem of cancer, and we need a major advance.'**

Prevention 'is going to involve everybody over their whole lives,' he declares. 'It's going to involve cleaning up the workplace and the environment, it's going to involve changing our diets, and it's certainly going to be a bigger hassle and more expensive than our ideal treatments would be.'

The NCI is already nominally devoting more of its budget to prevention research than to treatment research."

Scientific American, January 1994

This is how the state of cancer therapy is looked upon in one of the richest countries of the world. The most important message of the article is the change of attitude which, admitting the failure of present-day cancer treatment, focuses on prevention hoping to find the way out in a therapeutic approach utterly different from those applied until now.

WHAT IS TO BE EXPECTED?

As it is clear from the above, in one of the most advanced countries of the world, where in the last 20 years 35 billion dollars have been spent for treating cancer patients every year and 24 billion have been spent for research despite all efforts, cancer death rates went up by 7 percent. Taking into consideration the fact that in the years to come the number of risk factors is expected to grow, no special clairvoyance is needed to predict that the frequency of tumorous diseases will increase.

What lies behind this?

One of the causes is that **the man of today has missed his goal. He has forgotten that human society and technological civilisation can exist only because Nature has, for several hundred million years, created and sustained the system which provided us with air, clean water, fertile soil, precipitation, filtered sunlight, etc.** While the laws of economy urge participants to further expand production, increase GDP, the amount of crop per hectare and export, we are only now beginning to recognise that all these go hand in hand with the destruction of Nature which ensures our conditions for life. (To be more exact, decades ago responsibly thinking scientists cautioned us but the system, by inertia, keeps going on without any change of direction.) Harmful components of human activity have, during the last decades, surpassed the level which can be tolerated by Nature without causing permanent damage. Air and our waters are polluted to an extent never experienced before; the size of the ozone hole is constantly growing; the ozone layer is getting thinner, enabling the greater intensity of harmful UV radiation; the temperature of seas has increased; the mass of ice cover is diminishing; the bulk of the population on Earth live in metropolises of several million inhabitants where huge floods of cars emit harmful gases. Converting these global processes to the level of cells and tumorous diseases, we see that **unfortunately, there is an ever growing threat that defects in cells will increase and that while replicating genetic programs, raising the probability of the appearance of tumour cells, which will result in an increase in the number of cancer patients world-wide.** (According to prognoses, 20 million new cancer patients are expected to be diagnosed by 2020.)

WHAT CAN YOU DO TO DECREASE THE RISK OF CANCER, DESPITE THE ABOVE UNFAVOURABLE TENDENCIES?

Looking at the incidence of cancer in various states of the United States of America, we find an intriguing correlation. The figure is nearly the same in all states but one where it is strikingly low – a mere 50 percent of that in other states. This state is Utah where, due to the severe restrictions of the Mormon religion, the number of smokers and of people consuming alcohol and coffee is low. From this very simple statistical fact the most important thing to do becomes obvious.

Specialists agree that in a "single" step, without the use of drugs or medicinal products, cancer mortality could be reduced by 50 percent. There is nothing else we should do "only" to stop consuming obviously carcinogenic stimulants, change our outdated nutrition habits and, in general, care for our health more. Unfortunately, it seems as if the move in this direction, for some reason, is very difficult.

What prevents us from accepting and applying these very simple to follow principles?

If we are capable of rational thinking and action, why do 50 to 60 percent of adults consequently and consistently destroy their health while all of us care for and cherish our material goods and objects of value? How indignant we get when somebody scratches our car but we do not even notice that our children, sitting in the back seat, inhale the smoke of our cigarettes!

There are several answers to why we behave so irrationally:

- We do not know that what we are doing is harmful.
- We do not believe that rules refer to us, too.
- We are aware but take the risk, rather, in exchange for "enjoyment".
- We cannot imagine that there is an other way of life.
- We think that if there will be some trouble, doctors will "fix" us just like mechanics fix our car.
- We have no strength to change.
- We do not believe in the correlation between e.g. smoking and the higher incidence of cancer, etc.

Numerous arguments could be found for why we do not behave rationally but we do not intend to moralise. In fact, we would like to call attention to the fact that **the appearance of a tumour has just as simple and rational causes as that the car will be scratched if we wash it with a wire brush. The only significant difference lies in the dimension of time. While in the example referring to our rational thinking, an erroneous decision is followed by an immediate and obvious consequence, years may elapse from the appearance of a tumour and the causes leading to it. This misleads people because the connection does not seem so obvious.**

The first signpost on the road to change calls our attention to accept the observations of scientists, doctors and epidemiologists and at least try not to increase our own chances for tumours to appear.

Of course, it is not only the individual that has to change but also the views of society, politics and economic policy, as these, in fact, can only change together. It is futile for the individual to do his best for a healthier living if the air he inhales is strongly polluted; if the water he drinks is of a poor quality; if his food is full of pesticides, stabilisers, preservatives, colour agents and other artificial supplements. We should strive for rationality at a social level, too. The contradiction would be the topic of another book, namely that it is the interest of the individual, of economic units and of the country to produce, sell and consume, as this ensures our daily living, while all these activities in actual fact are continuously diminishing our living space and also our chances of survival. Narrowing down this circle of problems to the issue of cancer: prognoses claim that more and more people will get the disease at an ever younger age if we cannot take effective measures to alter the present tendencies.

It would, of course, be good if through individual and social decisions the consumption of harmful stimulants, and by this, the number of cancer incidences were to decrease but what would be needed most is to re-think how we could use the achievements of technological and scientific development to create systems and human communities where the chances of survival would grow and the feared diseases of our age – cancer, cardiovascular diseases or AIDS – would draw back.

Innumerable factors determine whether we live our lives with or without a tumorous disease. It is like the correlation between a car and the style of the driver. To have a good car (a strong and healthy organism) gives us an opportunity to drive at a great speed on the highways (in life) but it is not indifferent how we drive that car. If we do it cautiously (keeping the rules) we can go far but if we do not take notice of the traffic signs we can easily get wrecked. With appropriate circumspection you can also go far even if you have a medium or weaker car, but just like a cautious driver can have a car accident, you get cancer even if you lead the healthiest way of life. There is, unfortunately, no guarantee against cancer but it is possible to reduce the risk of getting it.

WHY IS IT NECESSARY TO CHANGE THE PRESENT PRACTICE?

The most important argument, naturally, is to avoid a premature death. This argument should in itself be sufficient to suppress all the others. Is there anything more important than to be able to bring up our children, to help them start their lives; is there a greater joy than to watch, as grandparents, how our grandchildren grow up while living our own lives in good health? According to British statistical data, half of today's 20-year-old smokers will die as a consequence of smoking and every second of them in their prime years. They will not live to see their children grow up. And these data are valid for every country in the world.

One should not neglect the economic consequences of being ill either. Everybody should know that a disease is an expensive thing for both patient and society. It is not true that smokers, through high taxes, contribute to the budget of a country, because five times the sum of taxes is spent on treating diseases connected directly or indirectly with smoking. Who would want to make a business of getting one thousand dollars for paying five thousand in exchange? And this is exactly what is happening now, only not with thousands but with hundreds of billions.

We can get richer not only by saving more money but also by trying to avoid certain expenditures with foresight and attention.

The man in the street only knows that it means a problem for the National Health Insurance to pay for the patient's medicine, that subvention is decreasing and the price of drugs is getting higher. All these happen in spite of the fact that throughout our life we give nearly half of our income to the Insurance Fund.

The appearance of newer cytostatic drugs (drugs used in cancer therapy) at the same time also entails a rise in prices. The development of these drugs consumes vast sums (USD 300 to 500 million) which, in turn, are added to the price of the manufacturer. The cost of an anti-cancer drug enough for 6 months can go up to 2–3 million forints (USD 6–10,000). It is easy to see that the treatment of a thousand patients means an expenditure of 2–3 billions (USD 6–10,000 million). But it is not a mere thousand patients that have to be treated – in Hungary 50 thousand new patients are diagnosed each year with cancer and the total number of tumour patients is somewhere between 150 and 200 thousand.

Even for the developed countries, the high-level patient-care means heavier burdens. It is in the interest of all countries to minimise the rate of cancer incidence and to find and apply the cheapest and most effective treatment procedures. Otherwise it might easily happen that the country's inhabitants will fall into two groups: the sick and those healthy who work to provide material cover for treating the sick. As a matter of fact, in a way this is what is happening today. The safest way to decrease these expenses would be to be able to cure more cheaply and to minimize the cancer rate.

The application of the procedure based on deuterium depletion in cancer therapy and prevention may contribute to the achievement of the above goals.

THE ESSENCE OF THE PROCEDURE BASED ON DEUTERIUM DEPLETION

WHAT IS DEUTERIUM?

Deuterium is the simplest element, a variation of hydrogen (i.e. its stable, not radiating isotope). The nucleus of deuterium contains a proton just like that of hydrogen but it also has a neutron of an equal mass *(Figure I.7)*. Thus the mass of deuterium is the double that of hydrogen. The significance of the above lies in the fact that the 100 percent mass difference leads to differences in the physical and chemical behaviour between the two isotopes, a phenomenon called isotope-effect.

WHAT DOES PPM MEAN?

When defining the deuterium content of Dd-water (deuterium depleted water), it is convenient to use the ppm unit. In our case, "ppm" (parts per million) shows, how many of one million hydrogen atoms can be considered deuterium atoms *(Figure I.7)* or how many heavy water molecules (D_2O) there are in one million water molecules.

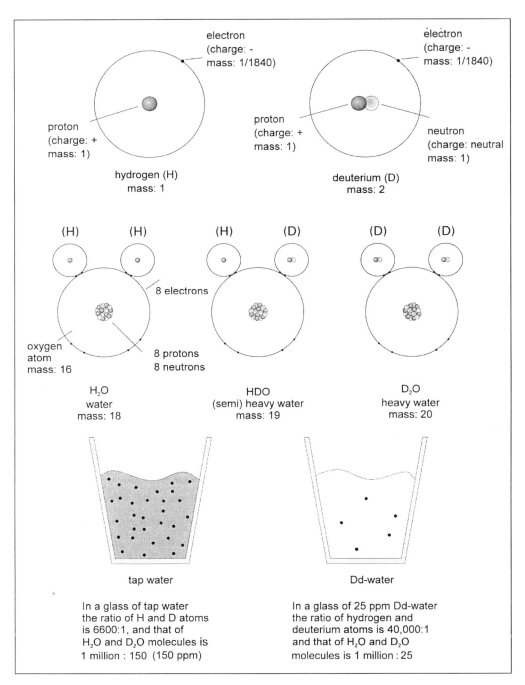

Figure I.7 *Hydrogen is the simplest element, it consists of a proton of positive charge and an electron of negative charge (atom mass: 1). The nucleus of deuterium in addition to proton has a neutron of the same mass but of neutral charge, this explains the 100 percent mass difference between the two elements. Taking into consideration the H–D isotope pair, in a glass of water there are simultaneously three kinds of water molecules. Producing deuterium depleted water, the H/D ratio is considerably shifted in favour of hydrogen.*

In our climatic zone, the D-concentration of natural water is around 150 ppm. Our aim was to significantly decrease the D-concentration of water compared to the natural value. The lower the D-concentration of water, the lower its D-content expressed in ppm, i.e. the greater the deviation from the D-content of natural water.

The first scientific experiments were carried out with water of 30–50 ppm concentration from which, at that time, we could produce a daily 20–30 grams only. Following this, we succeeded in producing a daily 20–30 litres of 100–110 ppm water. Later, in industrial circumstances, 4–5 tons of 85–90 ppm water were produced, while today we dispose of tons of 25 ppm Dd-water.

The D-concentration of the drinking water *Vitaqua,* marketed in 1994, was 130 ppm, whereas human clinical trials are presently being carried out with 90–95 ppm Dd-water. The presently available drinking water *Preventa-105* is of a 105 ppm D-concentration.

It should be known that by adding natural water to water of a lower D-concentration, any D-content can be produced (between the D-concentration values of the two water components) depending on the ratio of mixing.

For example: if we add 1 litre of 150 ppm natural water to 1 litre of 105 ppm water, the D-concentration of the mixture will be 127.5 ppm. (A method of exact calculation will follow in detail.)

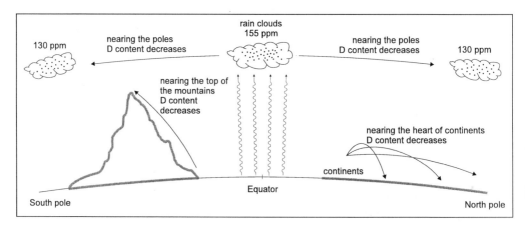

Figure I.8 *The D-concentration of precipitation measured at different points of the Earth varies, which can be explained by the differences in the behaviour of H_2O, HDO and D_2O molecules. In the Equator zone the D-concentration of the water evaporating from the oceans is close to the D-concentration of the oceans (154–155 ppm) but as cloud systems stream and lose their water content in the form of snow and rainfalls, their D-content decreases, because HDO and D_2O molecules "fall out" from the clouds more frequently than the ratio of occurrence. Nearing the poles, the clouds gradually lose their heavy water contents also when they meet with high mountains (on the top of the mountains the heavy water content is lower than at the foot of the mountain). Similarly the heavy water content of the precipitation in the heart of the continent is lower than in areas near to the oceans.*

IN WHAT QUANTITY IS DEUTERIUM PRESENT IN NATURE?

The D-content of living organisms on earth is basically defined by the D-content of ocean waters and the evaporating precipitation in the form of rain and snow. Measuring the range of the D-content of precipitation in 100 different points of the Earth, it can generally be stated that the D-content of precipitation decreases when nearing the poles from the Equator and also from the oceans towards the heart of the continents, as well as in proportion to height above sea level (*Figure I.8*).

In our climatic zone the D-content of surface waters is about 150 ppm (with a minimal fluctuation), i.e. in a million of hydrogen atoms, 150 deuterium atoms can be found as opposed to values of 155 ppm in the equator zone or of 135–140 ppm in the northern part of Canada.

It can be stated, in general, that there is but one deuterium in every 6,600 hydrogen atoms. Its importance can, naturally, be belittled by saying that if the occurrence of deuterium – in contrast to hydrogen – is so rare, it surely is of no significance whatever, it cannot play a serious role. This approach has been characteristic for the last 60 years, i.e. since we know of the existence of deuterium.

Taking into consideration that in surface water (oceans, seas, rivers) the concentration of deuterium is 150–155 ppm, it follows that in living organisms – which are not made up of water only – D-concentration varies between 110 and 120 ppm. This value becomes significant if we compare it to the concentration in human blood of other vital elements, such as calcium or magnesium, where deuterium is present in an approximately 6 times greater quantity than calcium and in a 10 times greater concentration than magnesium!

Such an element has been neglected by science for the past decades.

THE DISCOVERY

It was a simple question that led to the discovery: does naturally occurring deuterium have a role in biological processes?

The simplest way to answer the question is to investigate whether deuterium depletion has an effect on the biological processes of cells and of living organisms (those of plants, animals, and humans).

Investigations carried out during the past 9 years have proved in many ways that as a consequence of deuterium depletion, many processes occur differently in various biological systems. These differences show that cells are able to perceive the decrease in D-concentration and this fact triggers various processes in them.

We may conclude from the above observations that during its evolution of many billion years, Life has made use of the possibility that hydrogen has a variant of mass number two, i.e. deuterium, and has built a system of regulation on this.

We have, thus, not discovered a miraculous anti-cancer drug but a hitherto unknown regulating system present in cells. Its significance in tumour therapy is that experiments revealed that **by deuterium depletion the metabolism of tumour cells, i.e. their regulating system can be disturbed, which results in the destruction of tumour cells.**

WHAT LIES BEHIND THE ANTI-TUMOUR EFFECT OF DEUTERIUM DEPLETION?

Mention has already been made of the complex process of multiple steps, which can finally lead to the appearance of tumours in the human body. The question justly arises as to where and how deuterium depletion can have a positive influence on the outcome of this complex process. It is also clear that the appearance of tumours can be attributed to two main causes: one being an increase in the number of genetic defects, the other the increase of genetically defective (tumorous) cells. The former results in the greater probability of the tumorous nature of a given cell, the latter leads to the fact that among cells of a tumorous nature increasing in number there will probably be one or more tumour cells which, due to newer genetic defects will be able to evade or bridge over the obstacle preventing tumour growth, thus rendering cell proliferation to become utterly uncontrollable.

In connection with the "anti-cancer" effect of deuterium depletion we do not claim that by consuming Dd-water the number of mutations is decreased, although there are numerous arguments for this as well. (It is proven by experiments that in a medium with a higher than natural D-concentration, the number of mutations increases, so there is a good reason to presume that the presence of deuterium can be considered to be a mutation factor, thus, low D-concentration could as well, decrease the frequency of the incidence of mutations.)

Of more importance is to analyse the effect of deuterium depletion on tumour cells. Based on our observations so far, we can state that with the consumption of Dd-water the lowering of D-concentration may inhibit tumour growth and consequently may lead to partial or complete tumour regression.

This effect of Dd-water is reinforced by several data:
1) Cell proliferation was inhibited in Dd-water treated in vitro culture and, within our experimental system, deuterium depletion induced apoptosis.
2) In case of tumour transplants in mice, cats and dogs with spontaneous tumours as well as in cancer patients, the size of tumours stagnated or decreased as a result of Dd-water consumption. Experiments with mice

also verified that deuterium depletion inhibited the proliferation of tumour cells and resulted in their destruction inside the tumour: the process resulted in a decrease of the tumour volume.

3) One indirect proof of the presence of certain tumours is the appearance of specific proteins (tumour markers) in the blood. After the consumption of Dd-water (in some cases after a temporary increase), the value of tumour markers may decrease. This decrease is connected with the decrease of tumour volume, which also reinforces the anti-cancer effect of deuterium depletion.

4) Several examples have shown that in total remission (when in treated cancer patients there is no proof of the presence of tumour cells), in some cases there was no relapse for years when the patient had been drinking Dd-water, but when the consumption was stopped, the tumour appeared again within several months. This reveals that Dd-water helped these patients to maintain the balance of the organism and to inhibit tumour growth.

5) In developed countries numerous investigations were followed for more than 10 years, with the participation of many thousand patients to prove the connection between nutrition or drug usage, and the more or less frequent occurrence of certain diseases. One of these experiments has convincingly shown that among patients who regularly take NSAIDs (non-steroid anti-inflammatory drugs such as e.g. Aspirin) certain cancers (lung, colon) occur less frequently. Basic research has also proved that this effect is due to the fact that some drugs inhibit the gene COX-2 that has a role in prostaglandin synthesis. Our research has likewise shown that deuterium depletion inhibits COX-2 activity.

6) During the past months we have proved successfully that deuterium depletion inhibits the activity of those genes whose role in cancer growth has been known to scientists for years.

Our basic research results are in harmony with the work of researchers abroad and prove – both macroscopically and on a molecular level – the anti-tumour effect of deuterium depletion. We suppose that the presence of deuterium is vital for cell proliferation and that cell division is triggered by the change of the D/H ratio (the ratio of D temporarily increases in contrast to that of H). When the patient consumes water of a normal D-content, it is no problem for tumour cells to ensure the D/H ratio needed for division. When, however, we decrease the D-concentration of the body by deuterium depletion, either the conditions for cell division are not given, or the tumour cell can only achieve this ratio considerably later. We, thus, deprive tumour cells of a most important factor: the possibility to ensure conditions for cell division.

As shown earlier, cells are able to adapt to several external effects and with time among the numerous cells there will always be some which are able to achieve – through a genetic change – an advantage over other cells and, by outgrowing their surroundings, to overcome the obstacle. These facts make deuterium depletion an effective tool in tumour therapy but they also set the limits of application. **Deuterium depletion might be a successful means of destroying tumour cells because, in the adaptation process, healthy cells have an advantage over defective ones. This means that the unit decrease of D-concentration induces a stress in both healthy and tumour cells but healthy ones quickly adapt to the lower D-concentration whereas tumour cells are unable to do so or, and even then only, much slower. If, in the meantime, D-concentration continuously decreases, the tumour cell is again subject to a newer stress, although it could not even manage the previous one. With the appropriate dosage of Dd-water, this process may result in the destruction of all cells constituting the tumour.**

The constraints of the treatment are that the tumour, even if slowly, nevertheless is able to adapt to the new situation, the bigger the tumour and the more tumour cells the patient has, the easier. Unfortunately, there is a chance that after a certain time cells able to adjust to D-depletion will appear, resulting in further tumour growth. One of the directions of our drug development is to further enhance the efficacy of D-depletion by adding other materials. Another option is to give Dd-water at the earliest possible stage together with other treatments even if the patient, due to surgical or other treatments, is tumour-free because in such cases the number of possible tumour cells is minimal, giving us the greatest chance to destroy them.

THE POSSIBLE PREVENTIVE ROLE
OF DEUTERIUM DEPLETION

Because of the relatively short period (6.5 years) and low number of patients (1000) at our disposal so far, we cannot present direct data as to prevention. However, on the basis of results achieved during the past years we are able to draw certain consequences as to the best and most economical possible pattern of application for using Dd-water as a means of prevention. Based on our results we might suppose that Dd-water causes the destruction of potential tumours below the level of demonstrability if such an effect is unanimous in the case of tumours of a size belonging to the range possible to follow. Thus, Dd-water might be a harmless means of prevention (as opposed to when healthy women belonging to a high risk factor group for mammal cancer are offered cytostatic drugs) which could make it possible for us to interfere with the multi-level process demonstrated in *Figure I. 6* not only in its last phase but at a much earlier stage.

According to our present-day knowledge, the continuous consumption of Dd-water for prevention is not needed. It is enough to drink it as a preventive cure for 1–2 months a year. Our recommendation is to drink 0.5–0.7 litres of 105 ppm Dd-water for 2–3 weeks, then 1–1.4 litres for a further 2–3 weeks followed by 0.5–0.7 litres for a final 2–3 weeks *(Table II.20)*. From the aspect of Dd-water consumption the whole of the population can be divided into two main groups. To the first belong the healthy (Group H) and to the second those in whom some tumorous disease has already been diagnosed (Group C – cancer). The protocol of application for Group H is given below; the rest is dealt with in detail in the part for doctors.

Group H

To this group belongs the majority of the population, i.e. people who are not even suspected of having tumorous diseases. If we look at the age grouping of cancer mortalities *(Figure I.3)*, it becomes clear that the number of cancer incidences begins to increase after the age of 40. From the point of view of prevention and taking other aspects into consideration, we have made further sub-groups within *Group H*.

H/I

The population under 40, belonging to a low risk factor group, with no hereditary predisposition background.

This group has the lowest cancer incidence. People who live in unpolluted areas, do not smoke, consume healthy food, are not exposed to carcinogenic materials at their workplaces and have no family history of cancer, belong to this group. For them it is sufficient to repeat the protocol once every 3–4 years.

H/II

People under 40 who belong in one or more respects to a higher risk factor group.

It is clear that this group includes those who live in polluted industrial areas, and by their way of living (smoking, alcohol consumption and inappropriate nutrition) increase the probability of tumour growth or, even if these factors are not present, the genetic predisposition for tumour growth is demonstrable. For them, a Dd-water protocol every 2–3 years is recommended.

H/III

The population above 40, regardless of whether they belong to a low or high risk factor group. Those in this group should repeat the protocol once in every 1–1.5 years.

When determining the recommended daily dosage, we took a patient with an average of 55–70 kg (122–155 lb.) of body weight. (In case of a greater weight the dose should be increased rationally.) The daily dose of 0.5–0.7 litres was considered to be half of the average daily water consumption. (Note: Mixed with normal water in a 50:50 percent ratio, from 105 ppm Dd-water a drinking water of 127.5 ppm can be obtained. The protocol can also be realised by drinking a daily 1–1.4 L of Dd-water, which, in the first and last 2-3 weeks of the protocol is produced by a double thinning of 105-ppm Dd-water.)

CHAPTER TWO

To the Patient

INTRODUCTION

It is important to stress right at the beginning that Dd-water is not a wonder drug but it can be stated with great certainty that deuterium depletion may open a new era in tumour therapy. This is not merely because of the efficacy of the procedure but because of the revolutionarily new therapeutic principle by which we approach treatment and because a great therapeutic advantage can be obtained with minimal risk. We estimate that with the co-operation of specialists in question, the cancer mortality rate could, within 2–3 years, be decreased by up to 20–30 percent. With intense research and further clinical trials this ratio could be even further improved.

We know that tumour patients are in a very difficult situation for several reasons. Apart from suffering from the symptoms of the disease, a number of circumstances further aggravate living together with the disease: the feeling of helplessness, fear and ambiguity; the lack of information as to treatment; the sometimes tiring and exhausting waiting before and during medical examinations; an enhanced psychical load before control tests; the frequently experienced lack of compassion; the concomitant material burdens.

The primary aim of this book is to make people aware of the fact that they can and should do a lot in order to avoid cancer. From the intracellular processes, however, it also becomes clear that it is only possible to decrease probability – there will always be tumour patients, thus the challenge to treat them at the highest level possible will always be present in society. With the information on deuterium depletion we not only intend to present a new procedure to prevent cancer, to decrease its incidence and to effectively treat it but we wish to reach an even more complex goal, namely that society unite to solve the problems of cancer. Everyone should be aware that a lot can be done to defeat cancer and that it is absolutely necessary to do so. For those, however, who have already been diagnosed with the disease, a high-level humane medical care should be ensured in which the most effective treatment is available for everyone.

This chapter was written, in the first place, for those who have already been diagnosed with the disease. We have summed up for them the information we consider to be of utmost importance hoping they find it useful and with the spirit of our book and the new procedure we can contribute to the victory over cancer.

IMPORTANT ASPECTS IN ACHIEVING OUR GOAL

By using deuterium depleted water we wish to achieve a simple goal, the need to decrease the amount of deuterium in the body. This can be achieved with the consumption of Dd-water because it "mixes" with the water content of the body thus lowering the D-concentration of the body to that before the consumption of Dd-water. If this process is repeated day by day, the D-concentration of the body will continuously decrease and this might result in the destruction of tumour cells. To achieve this, a dosage of optimal concentration and amount is necessary.

APPROPRIATE DOSAGE

The measure of the decrease in D-concentration is determined by two main factors: a) the lower the D-concentration of the water and b) the smaller the water volume it is mixed with, the greater the decrease in the concentration on consumption of a volume unit of Dd-water.

To realise the aim of the treatment the following should be kept in mind: a) the patient should get a product of an appropriate D-concentration; b) the daily amount should reach the necessary level; c) the effect of Dd-water should not be spoiled by consuming great amounts of water of a normal D-content.

The decrease in D-concentration will not result in the desired effect: if the patient consumes Dd-water of a concentration higher than needed or if, beside Dd-water, he consumes normal beverages (water, soft drinks, tea, milk, beer, wine) in excess amounts. Beside the recommended amount of Dd-water, the patient can, of course, consume a serving of soup made with normal water, or eat vegetables and fruit but the consumption of considerable amounts of normal water is not recommended. To avoid diluting the D-concentration at meals, when drinking water or when eating fruit, we recommend the intake of 0.5–1 dl undiluted Dd-water of a 25 ppm D-content during meals.

HOW TO DIVIDE THE DOSE DURING THE DAY?

It is recommended to consume the necessary daily dose in 1.5–2 dl portions. It is good to begin the day with a glass of Dd-water and drink the rest during the day at equal intervals. The last portion should be consumed just before going to bed. We also recommend that Dd-water consumption should precede meals by 10–15 minutes and during meals 0.5–1 dl of undiluted (25ppm) Dd-water should be taken beside the daily amount.

WHAT ARE THE MAIN ASPECTS OF NUTRITION?

According to our observations, Dd-water is less effective if the patient follows a strict vegetarian diet. From among meat products we recommend the consumption of white meats (poultry, fish) in the first place and from dairy products yoghurt is the most favourable. As to fruits, the kinds grown in Hungary are preferred to tropical ones. *Figure I.8* shows that nearing the Equator, the D-concentration of the precipitation is higher, and this has an influence on the D-content of fruit grown in that region. This, of course, does not mean that the patient is not allowed to eat a banana or an orange occasionally but their daily intake should be limited.

Mention must be made of the fact that the D-concentration of fruit concentrates may be higher than that of natural drinking water. We do not think that for the majority of people this bears any significance but for patients with whom we wish to achieve a defined daily decrease in D-concentration, the D-content of other food beside Dd-water is not indifferent.

Any product which contains deuterium in a greater concentration, diminishes the effect of Dd-water, we therefore recommend to limit their consumption or to totally abstain from it.

CAN DD-WATER BE BOILED?

Deuterium depleted water is bottled enriched with carbon dioxide. This might be disturbing for the patient especially in the case of tumours of the oral cavity or those of the stomach. To avoid unpleasant effects, Dd-water can be boiled to make tea or soup with it, boiling in itself does not change the compound or efficacy of the product. It is nevertheless recommended to boil Dd-water for a minimum amount of time only, and under a lid to avoid getting in direct contact with air or its vapour content.

WHAT OTHER SUPPLEMENTARY TREATMENTS ARE RECOMMENDED?

Patients and their families are in a very difficult situation when they want to find their way in the maze of possible alternative medications and treatments. There are many approaches, several treatments and cures available at the same time and it is hard to decide which one to try from among the possibilities. The best is not to amass various treatments and not to use erratically one drug after the other but to choose two or three convincing procedures. Another issue of importance is how persistently these treatments are followed. Errors can be made in both ways: one mistake is when the ineffectiveness of some product has already been proved the patient nevertheless keeps to the given treatment.

The other is when the treatment is stopped before time, when it cannot as yet be expected to show a positive effect. There is no absolute truth concerning these questions, therefore it cannot be said that one procedure should be continued this long and the other that long. It is important, however, to choose a procedure with due circumspection and to apply it consistently within the limits of rationality.

Many patients have tried the so-called Béres-drops along with Dd-water, and the drugs Culevit and, later, Avemar. It is impossible to tell which product had a role and if so how great a role in the improvement of a patient and also, what would have happened had he applied only one or the other. What can be justified on the basis of observations is that the products have reinforced rather than weakened each other's effect.

We consider it a guiding principle that Dd-water should be applied alongside with conventional treatment.

THE EFFECTS OF Dd-WATER CONSUMPTION

Dd-water has, for the last years, been consumed primarily by tumour patients and only by a few who had no detectable tumours. It can be stated in general that in the latter the consumption of Dd-water had no perceivably expressed effect. Cancer patients, however, perceived the effect of Dd-water. Just as an inflamed tooth has its symptoms, when tumour cells start to decline massively, this, naturally, also has certain signs. Symptoms experienced by tumour patients can be explained by the interaction of Dd-water and the tumour, as a result of which an inflammatory process takes place in the tumour and its surroundings. On the basis of observations, accounts, and tests it can be stated that the accompanying symptoms are the consequences of the Dd-water–tumour interaction. There are generally observed and differing phenomena in various tumours. All these can be attributed to the differences in sensitivity, localisation and size of various tumours as well as dosage and a number of other factors.

THE MOST FREQUENTLY OBSERVED ACCOMPANYING PHENOMENA

GENERAL WEAKNESS, DESPONDENCY, SOMNOLENCE

These accompanying symptoms were experienced by most patients. They appeared generally some weeks after starting Dd-water consumption and lasted for various periods. It is important to note that when the dose is increased, despondency and somnolence appear again. If these symptoms disappear and the tumour is still present, the dose must be increased further. Symptoms do not appear if the size of the tumour is small (1–2 cm in diameter) as this does not perceivably burden the patient. (The phenomenon was observed in the case of tumorous cats and dogs which had been lying or sleeping for weeks). Weakness and somnolence are attributed to the necrosis of tumour mass and the reactions connected with it.

FLUSH, OCCASIONAL HIGH TEMPERATURE

Sudden attacks of fever were experienced in an advanced state and in the case of a great tumour mass. It is a known fact that having reached a certain size, the tumour, in some cases, undergoes spontaneous necrosis which, in a late phase of the disease, is accompanied by occasional high temperature. The phenomenon was observed during the application of Dd-water and could be attributed to necrosis affecting a greater mass of cells. Flushing of the face or local skin blush were frequently observed.

TEMPORARY INCREASE OF PAIN

The beginning of Dd-water consumption does not automatically and immediately result in the subjective well-being of the patients. The organism has to subdue the aggressive growth of the tumours. Unfortunately, as the consequence of treatment there might be a temporary increase in pain as well. This usually occurs in the case of bone metastasis and, to a lesser degree, with tumours of the viscera. We advise the patients to consult their doctors and to bridge this period with the help of prescribed painkillers.

ALLEVIATION AND CESSATION OF PAIN

It is difficult to foretell whether the alleviation or cessation of pain in a given patient is preceded by the increase of pain, but alleviation or cessation of pain is one of the surest signs of improvement. The cessation of pain is explained by the improvement of the original disease.

SWELLING AND SOFTENING OF THE TUMOUR AREA

In the case of tumours near the skin surface, the size of the tumour often increased but also became softer. Continuation of the treatment in a certain time resulted in a considerable regression (decrease in size) of the tumour. The temporary "growth" might be due to inflammatory reactions accompanying Dd-water consumption.

WARMING UP OF THE TUMOUR AREA

Patients on several occasions experience that tumours close to skin surfaces have become perceivably warmer. This observation correlates with former observations. In such cases the cooling of tumours is recommended as this might enhance the effect of Dd-water.

TINGLING, CRAWLY FEELING INSIDE THE TUMOUR
These feelings are also accompanying phenomena of processes occurring in the tumour.

BLEEDING IN THE CASE OF TUMOURS OF THE BLADDER AND RECTUM
The location of the tumour determines the consequences of tumour necrosis. In the case of the two above tumours it frequently happens that necrotic pieces of tissue become detached from the tumour and leave the body, accompanied by minor bleeding.

INCREASE OF APPETITE AND STRENGTH
Following an initial and transitory worsening, up to 50–60 percent of the patients felt better physically, their appetite grew and their strength improved.

WEIGHT GAIN
Beside the above changes we often experienced weight gain with patients. In some cases (several out of a thousand) the weight gain happened very quickly and reached up to 15 to 20 kilograms. In our view this was possible not merely by the positive effect exerted on the basic disease but also by other physiological changes evoked by Dd-water, and connected to the secondary effect exerted on the thyroid gland.

STOMACH COMPLAINTS
Having consumed Dd-water for a longer time, some patients complained of stomach problems. In such cases we advised them not to drink Dd-water in its carbonated form but make tea with it. This usually was enough to eliminate the problem.

WOUND HEALING AFTER A STRONG DISCHARGE
(WITH OPEN, ULCEROUS TUMOURS)
In the case of ulcerous tumours following the consumption of Dd-water first a strong discharge appears. This can be interpreted as a positive sign proved by the fact that in certain cases, in the place of the one-time tumour, a "crater" remains, which gradually closes.

IMPROVEMENT OF THE GENERAL CONDITION
Some weeks or months after beginning the consumption of Dd-water, patients feel better and their stamina improves.

"BRICK POWDER" IN THE URINE
In several cases a brick powder-like material of reddish or rusty colour appeared in the urine which could be attributed to the necrosing mass of tumour. It is advisable to consult a doctor and test blood levels of uric acid.

DECREASE IN SIDE EFFECTS OF RADIATION OR CYTOSTATIC TREATMENT

Beyond its effect on tumours, Dd-water also helps conventional therapy. Blood tests of patients consuming Dd-water did not worsen in the usual manner and in some cases hair loss also stopped (even if it had previously been a concomitant of treatment) as did strong nausea. Dd-water consumption, but especially the use of skin ointment made with deuterium depleted water alleviated skin complaints that had appeared as a consequence of radiation therapy.

A MORE SEVERE COUGH (IN CASE OF LUNG CANCER)

In the case of tumours of the lung the effect of Dd-water manifests itself in the strengthening of the urge to cough. This, too, is in connection with the size and location of the regressive tumour area. Patients coughed up white sputum mostly in the mornings which, in some cases, was slimy and sticky. Following the cessation of the urge to cough, patients could breathe more easily, their lungs could be filled with more air. The mucus in some cases also showed haemorrhaged threads.

THE BASIC RULES OF SUPPLEMENTING WITH DEUTERIUM DEPLETED WATER

1. Having consulted the doctor the patient should consume Dd-water in recommended quantities, together and in harmony with conventional treatment. The present-day legal situation and our knowledge, for the time being make the consumption of Dd-water possible only together with the application of conventional treatments. As our knowledge widens, there surely will be cases when Dd-water can be used as an exclusive treatment. It is only then that we shall know which cytostatic drugs help the effect of Dd-water and which can entirely be substituted by deuterium depleted water.

2. The untimely interruption of Dd-water treatment is not recommended. This is the most important advice to follow. Dd-water displays its effect rather slowly, in several weeks or months. If, after a cure of several months, the patient interrupts the consumption of Dd-water, he gives for tumour cells a chance to re-start dividing beyond control. If, in the course of conventional treatment, the patient cannot consume any liquid for some days or a longer time because of severe nausea, it is advisable to start consuming Dd-water only after having finished cytostatic treatment.

It is advisable to continue DD-water consumption after having reached total remission. The duration of the cure is difficult to determine, as we do not have adequate experience. In general it is recommended to have some months break after a half to one year treatment.

3. The first break in the treatment should not last longer than 2 or 3 months.
Whether the treatment resulted in total remission, can be stated with great
certainty 5 to 6 years after the completion only. A patient cannot be considered
cured even if he has been free of symptoms or complaints for several years, thus
it cannot be excluded that when a patient stops drinking Dd-water, residual
groups of tumour cells will start to grow again. Therefore, after the first break
of some months, we recommend a protocol of 4 to 6 months again. This can
be followed by a longer break of 4–6 months to be continued with a protocol
of 2–3 months.

A detailed guide to the application of Dd-water is given in the part destined
for doctors.

Book Two

Just as a musician constantly looks for the most beautiful sounds, music in the Universe is there to be played.

Marlo Morgan: Mutant Message Down Under

Book Two

CHAPTER ONE

To the Researcher

INTRODUCTION

In this chapter we wish to turn, in the first place, to researchers and those who specialise in biology. Our aim is to summarise scientific results in connection with deuterium depletion; to give a brief account on what the role of naturally occurring deuterium can be in cell functioning and finally, to describe how our results are linked to the knowledge discovered by molecular biology during the past decades.

In the course of our work we have been looking for the answer to the question whether naturally occurring deuterium has any role in living organisms. Results have confirmed that deuterium plays a key role in the regulation of intracellular processes.

PUBLICATIONS ON THE BIOLOGICAL EFFECT OF Dd-WATER

1. Somlyai, G., Jancsó, G., Jákli, Gy., Vass, K., Barna, B., Lakics, V., and Gaál, T. (1993): Naturally occurring deuterium is essential for the normal growth rate of cells. *FEBS Lett.* **317**, 1-4.
2. Somlyai, G., Jancsó, G., Jákli, Gy., Laskay, G., Galbács, Z., Galbács, G., Kiss, A.S., and Berkényi, T. (1996): A csökkentett deutériumtartalmú víz biológiai hatása [The biological effect of deuterium depleted water]. *Természetgyógyászat* **10**, 29-32.
3. Berkényi, T., Somlyai, G., Jákli, Gy., and Jancsó, G. (1996): Csökkentett deutériumtartalmú (Dd) víz alkalmazása az állatgyógyászatban [The application of Dd-water in veterinary practice]. *Kisállatorvoslás* **3**, 114-115.
4. Somlyai, G., Laskay, G., Berkényi, T., Galbács, Z., Galbács, G., Kiss, S.A., Jákli, Gy., and Jancsó, G. (1997): Biologische Auswirkungen von Wasser mit vermindertem Deuteriumgehalt. *Erfahrungsheilkunde* **7**, 381-388.
5. Somlyai, G., Laskay, G., Berkényi, T., Jákli, Gy., and Jancsó, G. (1997): Naturally occurring deuterium may have a central role in cell signaling. In: Heys, J. R. and Melillo D. G. (eds.) *Synthesis and Applications of Isotopically Labeled Compounds.* John Wiley and Sons Ltd. 137-141.
6. Belea, A., Kiss, A. S., and Galbács, Z. (1997): Eltérő módon asszimiláló növények C3, C4 és CAM típusának meghatározási módszerei [Methods of defining types C3, C4, and CAM of plants. *Növénytermelés* **46**, 477-485.

7. Somlyai, G., Laskay, G., Berkényi, T., Galbács, Z., Kiss, G. A., Jákli, Gy., and Jancsó, G. (1998): The biological effects of deuterium-depleted water, a possible new tool in cancer therapy. *Z. Onkol./J. of Oncol.* **30**, 91-94.
8. Somlyai, G. (1998): Csökkentett deutériumtartalmú víz – Új lehetőség a daganatterápiában [Deuterium depleted water – A new possibility in cancer therapy]. *Komplementer Medicina* **2**, 6-9.
9. Somlyai, G. (2000): Új daganatellenes állatgyógyszer kutyák, macskák kezelésére [A new anti-tumor veterinary medicine for treating cats and dogs]. *KisállatPraxis* **4**, 24-28.
10. Gyöngyi, Z. and Somlyai, G. (1999): Deuterium depletion can decrease the expression of *c-myc*, *Ha-ras* and *p53* gene in carcinogen-treated mice. *In vivo* **14**, 437-439.
11. Laskay, G., Somlyai, G., Jancsó, G., Jákli, Gy. (2001): Reduced deuterium concentration of water stimulates O_2-uptake and electrogenic H^+-efflux in the aquatic macrophyte *Elodea Canadensis*. *Journal of Deuterium Sciences* (accepted for publication)

Access to the original publications in English, German and Hungarian: HYD Ltd. for Research and Development (www.hyd.hu).

LECTURES, SEMINARS, AND POSTERS IN CONNECTION WITH THE BIOLOGICAL EFFECT OF Dd-WATER

1. Arbeitsgemeinschaft Stabile Isotope E.V.
    Bayreuth, September 1992 (lecture)
2. Wissenschaftlicher Kongress der Gesellschaft für Biologische Krebsabwehr
    Heidelberg, June 1993 (lecture)
3. 46th Annual Symposium on Fundamental Cancer Research
    Houston, October 1993 (poster)
4. New aspects in molecular medicine 2
    Zagreb, November 1993 (lecture)
5. XVIth Meeting of the International Society of Pediatric Oncology
    Paris, September 1994 (lecture)
6. 3rd International Conference of Alternative Medicine
    Budapest, October 1994 (lecture)
7. Manitoba Cancer Foundation
    Winnipeg, January 1995 (seminar)
8. University of Illinois
    DeKalb, January 1995 (seminar)
9. 2st Congress of the Hungarian Oncology Society
    Pécs, November 1995 (lecture)
10. Reddy Memorial Hospital
    Montreal, October 1996 (seminar)

11. Congress of the American Association for Cancer Research: Cell signaling and cancer research
    Telf-Buchen, February 1997 (poster)
12. Sixth International Symposium on The Synthesis and Applications of Isotopes and Isotopically Labeled Compounds
    Philadelphia, September 1997 (lecture)
13. Hungarian Oncology Society, Transdanubian Section
    Pécs, May 1998 (lecture)
14. Academia Romana
    Bucharest, April 1999 (seminar)
15. Radioactivity – Water – and Life Workshop
    Bad-Salzuflen, April 1999 (lecture)
16. 4th Meeting of the Hungarian Biochemical Association's Molecular Biology Section
    Eger, May 1999 (poster)

RESULTS OF BASIC RESEARCH (1990–1999)

The following part sums up the results. In each case, the number of the publications listed in "Publications on the Biological Effect of Dd-water" with the results and the place of examinations is given in brackets. (Where there is no reference to any publication, results will be published in the future.)

1. Dd-water (30 ppm) inhibited the division of L_{929} fibroblast cells [(1), Országos Onkológiai Intézet/National Institute of Oncology (OOI), Budapest].
2. The inhibition exerted by Dd-water was more expressed if L_{929} cells were synchronised by serum deprivation in phase G_0/G_1, and hence transferred to deuterium depleted medium [(1), OOI, Budapest].
3. In nearly 60 percent of immunosuppressed mice with transplanted MDA and MCF-7 human breast tumours, drinking of Dd-water caused complete regression [(1), OOI, Budapest].
4. As an effect of Dd-water (90 ppm), the tumour growth was slowed down in mice with transplanted PC–3 human prostate tumours, and survival rate increased by 40 percent in the treated group [(6), Institute of Toxicology, Keszthely].
5. Dd-water (90 ppm) inhibited *in vitro* growth rate of the A4 cell line (murine haemopoietic cells) [(5, 6), Paterson Institute, Manchester, UK].
6. The intracellular pH of A4 cells was restored with a delay, after an artificial pHdecrease in deuterium depleted medium (90 ppm). The amyloide sensitive Na^+/H^+ antiport system, which has a key role in cell division, took place in the process [Paterson Institute].

7. Dd-water (110 ppm) induced apoptosis in mice with transplanted human PC–3 prostate tumours [(5, 6), Semmelweis Orvostudományi Egyetem/ Semmelweis Medical School (SOTE), Budapest].
8. Dd-water (90 ppm) induced apoptosis in interleukin-dependent A4 cell line *in vitro* [(5), Paterson Institute)].
9. Dd-water (90 ppm) activated the membrane transport process of the seaweed *Elodea Canadensis*, causing an increase in pH in cells, and a decrease in the extracellular area [(2, 4 and 11), József Attila Tudományegyetem/ József Attila University of Life Science (JATE), Szeged, Hungary].
10. Dd-water (90 ppm), simultaneously with the pH-change, hyperpolarised the cell membrane in the seaweed *Elodea Canadensis* [(11), JATE].
11. Dd-water (20, 60, 110 ppm), during germination, inhibited the sprouting of 10 different plant species [(2, 4, and 8), JATE].
12. In the case of rice, corn, wheat, and squash, a clear dose dependence was observed showing that the lower the D-concentration, the stronger the inhibition of sprouting [(2, 4, 8), JATE].
13. Dd-water inhibited DNA synthesis *in vitro*, in MCF–7 (human breast), PC–3 (human prostate), and M14 (human melanoma) cell lines [(5), Oncotech Incorporation, Irvine, USA].
14. Inhibition of DNA synthesis, in the case of MCF–7 (human breast), PC–3 (human prostate), and M14 (human melanoma) cell lines was higher and also lasted longer with cells synchronised in the phase G_0/G_1 *in vitro* [(5), Oncotech Incorporation].
15. In the inhibition of DNA synthesis, a dose dependence was clearly evident: the lower the D-concentration of the culture medium, the stronger the inhibition [Oncotech Incorporation].
16. The application of Dd-water (90 ppm) resulted in the total cure of a cat with lymphoid leucosis [(3), Alpha-Vet Veterinary Hospital, Székesfehérvár, Hungary].
17. Dd-water (90 ppm) resulted in significant tumour regression and healing in dogs with breast, rectum, and testis cancer [(2, 4, 5, 6, and 7), Alpha-Vet Veterinary Hospital].
18. Dd-water inhibited the proliferation of a HT–29 human colon tumour cell line *in vitro* but exerted a minimal effect only on healthy myometrial cells [SOTE].
19. The inhibitory effect of Dd-water was more expressed in the case of HT–29 cell line, too, when cells synchronised in phase G_0/G_1 were placed in a culture medium with Dd-water [SOTE].
20. Inhibition of DNA synthesis was observed as a result of deuterium depletion in case of tumours of the ovary and breast from newly operated tumour patients *in vitro* [Oncotech Incorporation].
21. During experiments carried out with human tumours, the inhibitory effect of Dd-water doubled (increased to 40 percent) in the experimental system,

and inhibition duration also grew when the decrease in D-concentration happened not in one step (150–20 ppm) but in 4–5 steps (150–60–55–51–46–42 ppm) in 2–3 days [Oncotech Incorporation].

22. Deuterium depletion (20 ppm) inhibited COX–2 gene expression in healthy myometrial cell line [St. Louis University Medical School, St. Louis, USA].
23. Deuterium depletion (20, 60 ppm) inhibited the expression of COX–2 gene in the case of HT–29 colon tumorous cell line [SOTE].
24. COX–2 gene inhibition correlated with the D-concentration of water, the lower the D-concentration, the more significant the measure of inhibition [SOTE].
25. The inhibitory effect of D-depletion on COX–2 also correlated with the prostaglandin concentration measured in cells. The more significant the inhibition of COX–2 expression, the lower the intracellular prostaglandin concentration [SOTE].
26. Deuterium depletion inhibited the expression of the *Ha-ras* and *c-myc* oncogenes as well as that of the *p53* tumour-suppressor gene in mice [(10), Pécsi Orvostudományi Egyetem/University Medical School (POTE), Pécs, Hungary].

THE BIOLOGICAL EFFECTS
OF DEUTERIUM DEPLETED WATER

DEUTERIUM

In nature, 3 isotopes of hydrogen occur: hydrogen (H – mass number 1), deuterium (D – mass number 2), and tritium (T – mass number 3). Deuterium is the stable, non-radiant isotope of hydrogen. It has been known for decades that, due to the mass difference between H and D, molecules with a D-content behave differently in chemical reactions [1, 2] (the numbers refer to those of the Bibliography at the end of the chapter). Thus, for example, if a chemical bond contains deuterium instead of hydrogen, during the chemical reaction this bond splits approximately 6 to 10 times more slowly. If hydrogen is replaced by deuterium not in the splitting bond but at a more distant point of the molecule, chemical reactions also slow down to a significant degree. The first method of heavy water (D_2O) production was based on the observation that, during the electrolysis of water, the decay rate of H_2O can be the multiple of that of D_2O. These so-called kinetic isotope effects allow a special insight into the mechanisms of chemical reactions and the replacement of hydrogen with deuterium is widely applied in chemical research. Nucleo-magnetic resonance experiments clearly reinforce that the presence of deuterium also has an effect on distant points of a given molecule, in this way greatly influencing the behaviour in chemical reactions of the molecule [3].

The chemical difference between deuterium and hydrogen is manifested in biological systems as well. In the past decades heavy water was used in great concentrations during experiments and it was observed that this significantly influenced processes occurring in the given biological system [4, 5, and 6]. In these experiments it was stated that e.g. the growth of tobacco plants slowed down to a great extent on the effect of increased D_2O concentration. A dramatic effect was observed in the case of the mould fungus *Aspergillus niger*. This, as indicated by its name, is a black coloured fungus, which, in a heavy water medium turns alabaster white which means that the fungus is unable to produce the pigment responsible for the black colour. Drinking D_2O worsened the overall blood-count picture in animals or, in the extreme case of drinking heavy water of more that 35 percent D-concentration, the animal (dog) perished. Research carried out with mice and rats showed that in mammals, as opposed to simpler living beings, hydrogen cannot completely be substituted by deuterium: animals endure an approximately 25 D-percent of body fluid, which can be reached by drinking heavy water of a 30 percent deuterium content.

The effect on living organisms of heavy water is not surprising if we take into consideration that a significant part of living organisms is made up of water and that heavy water differs from normal water (H_2O) in many aspects [2]. Its melting point is almost 4°C, and boiling point 1.5°C higher, its density is 10 percent higher and viscosity 25 percent more than that of natural water. All these support the generally accepted view according to which the structure of heavy water is "stronger" than that of normal water. Part of the hydrogen present in living organisms – H-atoms bound to oxygen, sulphur, and nitrogen – in a heavy water medium is quickly substituted by deuterium. In this way, hydrogen bonds responsible for the stability of proteins become deuterium bonds, but deuterium bonds are stronger than hydrogen bonds; this, in turn explains why proteins in heavy water are more stable when exposed to denaturalisation and conformational changes [5].

A common feature of experiments carried out till now and referring to the effect of deuterium on living organisms is that they did not assign any significance to naturally occurring deuterium, heavy water therefore was applied mainly in greatconcentrations.

In what concentrations does deuterium occur in nature?

The D-content of living organisms on Earth is defined basically by the D-content of ocean waters and the evaporating precipitation in the form of rain and snow. Measuring the range of D-content in several hundred points of the globe, it can generally be stated that the D-content of precipitation decreases when nearing the North and South poles from the Equator and from the oceans towards the continental inland, as well as in proportion to height above sea level [7]. This observation can be explained by the difference in steam pressure between H_2O and D_2O (or HDO) [2]. In our climatic zone the D-content of surface waters is about 150 ppm (with a minimal fluctuation) as opposed to values of 155 ppm in the Equator zone and of 135–140 ppm in the northern part of inland Canada.

If the 150 ppm value of D-concentration is expressed in terms of mmol/L concentration, we find that in natural waters the concentration of D_2O is 8.4 mmol/L, this – as deuterium is present in natural waters mainly in the form of HDO – in fact means a 16.8 mmol/L HDO concentration.

Taking into consideration that 60 percent of an adult's body is made up of water, and also regarding that organic compounds other than water also contain deuterium, we reckon that the D-concentration of our body is somewhere between 12 and 14 mmol/L. (The quantitative ratio of deuterium in the human body can be demonstrated by the following figures: in a person of 50 kg body weight, there are approximately 5 kg hydrogen and 1.5 g deuterium.) As a contrast, we mention that human blood contains approximately 2 mmol/L calcium, 1 mmol/L magnesium and 4 mmol/L potassium. Taking the above levels into consideration, the question as to what role deuterium plays in biological systems becomes obvious, knowing that the presence of certain elements (calcium, magnesium, potassium) which occur in much lower concentrations in human blood and within a narrow range of concentration are indispensable for biological functioning.

LIVING ORGANISMS CAN DIFFERENTIATE BETWEEN D AND H

Beyond physical fractionating, it has also been known for 20 years that in different biological systems and, within them, in certain molecules, the D/H ratio may significantly change in contrast with the D/H ratio of surrounding waters. Depending on e.g., whether the plant fixes carbon dioxide from the air *via* C3 or C4 pathways, the extent of the decrease in D-concentration may differ or, in plants belonging to the so-called CAM group, in certain circumstances D-enrichment (concentration) occurs [8]. This means that by defining the D-concentration of a plant, we can tell which of the above groups it belongs to from the aspect of photosynthesis. The refined and sensitive character of biological processes is also shown by the fact that in algae, during processes occurring by light, the cell differentiates between the two isotopes of hydrogen, whereas no such discrimination is made in dark [9].

From the point of view of our work, the recognition that the ATP-ase enzyme of yeast also discriminates between the two isotopes of hydrogen, is of major importance. The discrimination is manifested in the fact that the enzyme does not accept deuterium as a substrate, only hydrogen [10]. If with other living beings it will also be proved that such a selectivity exists, this will mean that during energy-gaining processes there is a possibility for the D/H ratio to significantly change in the cell or in certain organelles of the cell.

We suppose that in proteins participating in other hydrogen transport processes, such discrimination will also be revealed. It is obvious that the biological effect of deuterium has been studied intensively in the past decades but always in high concentrations, disregarding the amount of naturally occurring deuterium. The novelty of our approach lies in the fact that we examined whether deuterium depletion, a medium of decreased D-content, also induces a response in various biological systems.

During experiments carried out with water the D-content of which was decreased to various degrees, we found that naturally occurring deuterium plays a key role in the regulation of intracellular processes [11–17]. Based on our results, we might suppose that cells have a D–H metabolism which regulates the D/H ratio in cells and, through this, a number of other processes, too.

HOW TO PRODUCE DEUTERIUM DEPLETED WATER AND HOW TO MEASURE ITS D-CONTENT?

The production of water of a decreased deuterium content is based on the differences between the physical and chemical characteristics of normal water (H_2O) and heavy water (D_2O). When producing Dd-water, we made use of the fact that as a consequence of the difference in volatility, at the boiling point of normal water, the steam in equilibrium with the liquid contains approximately 2.5 percent less deuterium than the liquid phase. Repeating this evaporation – which in industrial quantities happens in distilling towers – the deuterium content of water may be decreased to preference.

Using this method we produce water of a deuterium content anywhere between 25 and 110 ppm. The other frequently used method is based on the fact that in the hydrogen gas developing during the electrolysis of water, deuterium concentration is 1/3 to 1/9 of that of water. Hydrogen thus gained oxidised (with oxygen) makes water with a depleted deuterium content. With this more expensive method, by repeated electrolysis, any D-content can be reached relatively easily.

The definition of the deuterium content of Dd-water was carried out in an infrared domain corresponding to the O–D oscillation of the HDO molecule containing a D-atom, by measuring the intensity of the adsorption summit at 4 μm wave length. After calibrating, using standard patterns of known D-content with the Foxboro Miran 1A CVF spectrophotometer, deuterium content can be defined with an exactness of ±3 ppm. Using the mass spectrometric technique an even greater exactness can be achieved.

RESULTS

Proliferation of cells in D-depleted culture medium

The first investigations in the regulation of cell division were carried out *in vitro* with animal cell cultures (L_{929}; MCF–7; A4; 416B) [15]. We found that in a medium of lower than natural D-content, cell division started with a delay of 5–10 hours, the medium of a decreased D-content, however, had but a minimal effect on subsequent growth. Results reveal that cells recognise the lack of deuterium but quickly adapt themselves to the new medium. This also shows that the stoppage or slowing down of cell division is caused by the difference in D-concentration. When the cells were exposed to a Dd-water medium for several hours, the difference between the control and the Dd-water culture medium decreased significantly. Our results were later reinforced by investigations carried out in the laboratories of Oncotech Incorporation, in Irvine, California. In the first series of experiments, scientists followed the incorporation of H^3-Thymidin into DNA after the change of medium in PC–3 (prostate), MCF–7 (breast) and M14 (melanoma) cell lines. Following deuterium depletion, inhibition was experienced in all cell lines. It is important to note that the sensitivity of cell lines was different, which manifested itself in the fact that, in the case of melanoma cell line, inhibition was maintained for 6 hours only, in prostate cell line for 24 hours; while in the case of breast cell line, inhibition lasted for 48 hours. In all cases, inhibition was stronger when cells synchronised in G_0/G_1 phase were exposed to Dd-water culture medium. The rate of inhibition was around 20 percent.

In the above cases cells were exposed to a decrease in D-concentration on one occasion only. For modelling the process in the human body which means a decrease in D-concentration lasting for several months and occurring day by day, we later had experiments carried out in the course of which the D-concentration of the culture medium was decreased not in a single step but in 2–5 steps (150–60–55–51–46–42 ppm) within 24–72 hours. These experiments were carried out *in vitro* with freshly removed human breast and ovary tumours. Results showed that the more the D-concentration was decreased, the greater the inhibitory effect, reaching, at the end of the third day, a value of 40 percent. This modelled experiment clearly demonstrates that with the continuous decrease of D-level, the inhibition of cell division can not only be maintained for a longer time but can also be increased.

We later demonstrated that the proliferation of HT–29 human colon tumour cell line could also be inhibited in Dd-water medium. We wish to stress that in a healthy myometrial cell line examined simultaneously, the Dd-water medium only slightly influenced the proliferation rate of cells.

EXPERIMENTS WITH MICE WITH TRANSPLANTED HUMAN TUMOURS

As *in vitro* experiments showed that in a Dd-water medium, cell proliferation starts with a delay and that by a multiple decrease in D-concentration inhibition can even be increased, we set off to examine the effect of Dd-water on the growth of human tumours transplanted in mice. The first experiment was carried out with two different breast cell lines (MDA, MCF–7). The animals were treated with 30-ppm Dd-water the day following transplantation. Three months later, from the 11 (5+6) tumorous animals in the two control groups only one survived, its tumour had spontaneously regressed, whereas in the treated groups, from 17 (9+8) tumorous animals, initial tumour growth stopped in 11, followed by total regression [11].

In two further independent investigations with mice, the effects of Dd-water on PC–3 human prostate tumour were examined. In the first experiment with prostate tumour the treatment of the animals (22 in the control group and 22 in the treated group) had started on the 32nd day after transplantation. The average tumour volumes were 1.2 cm^3 in both groups, which means an advanced state. Tumour growth was then followed individually in the animals and we found that in the control group the tumour size gradually increased in every animal except in one. In the treated group, the tumour volumes were found to increase continuously in only 8 of the animals, where the size of the tumour had been above the average. In 7 animals with near to average tumour volume, initial tumour regression was observed and it was only after 1–2 weeks that tumour volume started to increase. From 7 further animals with smaller tumours (0.1–0.2 cm^3) 3 exhibited complete tumour regression, while in 4 cases tumour growth was completely inhibited for 2–4 weeks [16]. Inhibition is well reflected by the fact that in the control group the average value of tumour volume was 11 cm^3 on the 74th day of treatment, in the group treated with Dd-water this number was a mere 4.3 cm^3.

In the second experiment we repeated the one carried out with PC–3 prostate tumour, this time investigating the anti-cancer effect of Dd-water on a cellular level. Following transplantation, both groups were given normal water for 18 days to allow the tumour to develop. After this, the treated group consumed Dd-water for 2 weeks, and then all animals were killed. The inhibitory effect of Dd-water was evident macroscopically, the average tumour volume was 40 percent smaller, but of more importance was the fact that the tumours in the treated group were almost regular, ball-shaped ones, whereas in the control group they were grown diffusely. Microscopic investigation of the frequency of cells in mitosis (dividing) and apoptosis (dying) revealed the following results [15, 16]:

Table II. 1 *Deuterium depletion inhibited cell division and triggered apoptosis* in vivo

Control group	Treated group
Mitosis : Apoptosis	Mitosis : Apoptosis
3.6 : 1	1,5 : 3

These investigations corroborated the finding that D-depletion causes cell death.

INVESTIGATION OF THE ANTI-TUMOUR EFFECT OF DEUTERIUM DEPLETION IN DOGS AND CATS

Results are presented through the description of three actual cases with the help of a summarising table *(Table II. 2)*

Case history 1:
A cat in a very poor condition was taken to the veterinary hospital. It was weak and had no appetite (body mass: 4 kg). Veterinary examination stated an apathetic animal uninterested in its surroundings, with medium degree anaemia and a palpable swelling of lymphatic glands. During the physical examination of the abdominal cavity, a growth of 2 x 3 cm in size was palpable, corroborated by X-ray imaging. During a consequent diagnostic laparotomy, the significant swelling of the mesentery lymphatic gland and a moderate swelling of all the other glands was found. One lymphatic gland was surgically removed, and pathological examination corroborated lymphoid leucosis.

Table II. 2 *The effect of Dd-water on animal (dog, cat) tumours*

Tumour type	Cured	Improving	No change	Worsening
Dogs				
Breast	2	7	-	-
Epithelioma	2	-	-	-
Diffuse seminoma	-	-	1	-
Rectum	-	1	-	-
Melanoma	-	-	-	3
Cats				
Lymphoid leucosis	1	3	-	-
Total	**5**	**11**	**1**	**3**

Because of the poor overall condition of the cat, in the first days Dd-water was given 5 times a day, orally with a syringe. In the seventh day of the treatment, the cat could drink by itself, and became lively. From this time on, its condition improved day by day, it became interested in its surroundings and in some days it could eat unaided. After three weeks of hospitalisation, although the enlarged lymphatic glands were still palpable, there were no other symptoms related to the primary disease. At the end of the third month of the treatment, a further lymphatic gland was removed which still showed the earlier diagnosis. After a further three months, during the repeated pathological investigation, lymphatic hyperplasia was diagnosed.

Two years after the first diagnosis the cat was healthy, had a good appetite and a body mass of 9 kg.

Case history 2:

The first Dd-water treated dog had a mammary tumour, the size of which was 10 x 6 cm at the beginning of the treatment and, in addition, several smaller metastases of 2–3 cm in diameter were palpable. The pathologists diagnosed it hystologically as adenocarcinoma. The consumption of Dd-water for 1–2 months led to a complete regression of the smaller metastases which was accompanied by the decreasing volume of the primary tumour. After five months the size of the tumour was 6 x 4 cm, and after further four months 3 x 2 cm, when it was removed surgically.

Case history 3:

In the case of the dog with the tumour of the testis, the tumour was diagnosed as being diffuse seminoma. As an effect of Dd-water (90 ppm), tumour growth stopped and although there was no regression, the size of the tumour stagnated during the several months long treatment. Following a break in the treatment, the tumour started to grow again and repeatedly stagnated after the renewed consumption of Dd-water. We consider it possible that in case of a smaller tumour size (it had originally been 5 x 6 cm), or with the application of Dd-water of lower than 90 ppm D-content, a regression could have been reached.

Mention must be made of the fact that Dd-water of a 90 ppm D-content did not prove to be effective in the case of dogs with melanoma. Three such dogs have been treated so far with no genuine improvement. In the case of this tumour type, treatment should be tried with a VETERA-DDW-25 mixture of a 72 or 65 ppm concentration *(see Appendix)* and the dose should be increased within a month. Our observations are in harmony with the *in vitro* results of Oncotech Incorporated, where it was shown that the melanoma cell line had adapted to the culture media with a decreased D-content 6 hours after the exposition, whereas in the case of prostate and breast cell lines this happened after 24–48 hours only.

THE EFFECT OF DEUTERIUM DEPLETION ON COX–2 GENE EXPRESSION

Experiments carried out in the past years have revealed that cyclooxygenase enzymes and the prostaglandins (PGs) synthesised by them have an important role in the regulation of tumour growth and metastasis [18, 19, 20].

Prostaglandins are formed from arachidonic acid in reactions catalysed by phospholipase A_2 through cyclooxygenases. Latest research has revealed that the enzyme cyclooxygenase has two isoforms: *cyclooxygenase–1* (COX–1 or PGH–1) and *cyclooxigenase–2* (COX–2 or PGH–2). COX–1 is bound to the cell membrane and to the membrane of the endoplasmic reticulum. Chemically, it is a molecule of a haemo- or glycoprotein nature. It is produced continuously in almost every cell. Its primary function is to mediate the synthesis of prostaglandins needed for the normal functioning of the cells *("housekeeping" function)*. Its quantity is nearly constant. COX–2 is not found in inactive cells or in very small quantities only. Induced by growth factors, endotoxin, and cytokines, its expression grows dramatically (a 10 to 80-fold increase in activity). During inflammation and cell growth this happens at a very early stage. Its amino acid sequence is to 60 percent, and molecular mass to 100 percent identical with that of COX–1 (70 kDal). On the basis of the aforesaid, it is obvious that the two isoenzymes have different intracellular functions. Division and other inductive processes are generally connected to the activity and expression of COX–2. Several experiments have proved that tumours synthesise more

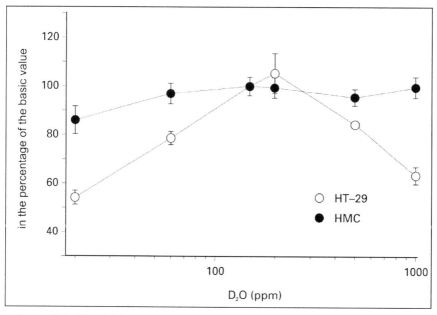

Figure II. 1 *The division of two cell lines: HT–29 human colon carcinoma and normal primary myometrial cells (HMC) in culture media with different D-concentrations*

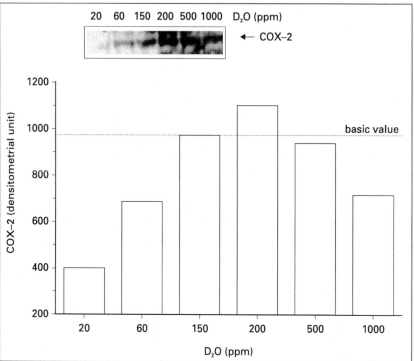

Figure II. 2
*The COX–2
gene expression
of HT–29
human colon
carcinoma cell
line in culture
media of
different
D-concentrations*

prostaglandin than healthy tissues [18]. It can, thus, be supposed that PGs play an essential role in tumour formation, in the regulation of tumour growth and in the development of metastases. This is confirmed by experiments where the inhibition of prostaglandin synthesis by non-steroid anti-inflammatory drugs (NSAIDs) caused tumour regression in the case of colon, breast, oesophagus, lung and oral cavity cancers [20]. The higher PG-concentration in tumour cells is the effect of the increased synthesis, for which the COX–2 isoform is responsible in the first place. Because of the possible cause-and-effect connection – between the overactivity of COX–2 and the higher PG-concentration on the one hand and the induction and metastasis of tumour on the other – it seems that targeted inhibition of COX–2 could be an effective approach in preventing cancer.

Experiments were carried out with normal primary human myometrial cells (HMCs) and with a tumorous cell line of human colon origin (adenocarcinoma HT–29). Human myometrial cells were obtained after the digestion by collagenase of uteruses removed surgically in hysterectomy. Both cells were cultured in RPMI media containing 10 percent foetal calf serum (FCS) in a carbon-dioxide thermostat (5 percent CO_2 : 95 percent air). When cells had almost filled the space in the culture dish (70 percent confluence value), we withdrew the serum for 24 hours. Following serum deprivation, the cells were given RPMI media prepared with deuterium depleted water, and after 24 hours the MTT (microculture tetrazolinum) assay was performed to determine cell proliferation.

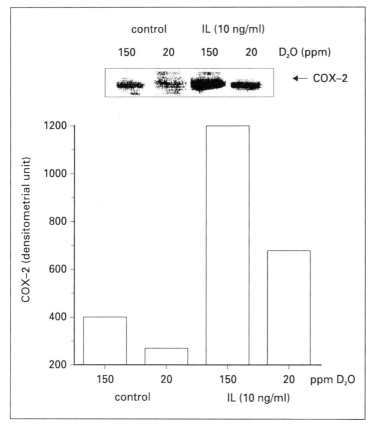

Figure II. 3 *The effect of Dd-water on COX–2 gene expression in cells both basic and stimulated with cytokine (IL-1) in myometrial cell lines*

Figure II.1 shows the division graphs of the two cell lines taken as a function of D-concentration. The figure shows well that the D-content of 20 ppm inhibited (in 45 percent) the proliferation of HT–29 cells. Inhibition was connected to deuterium concentration. The degree of inhibition decreased continuously and at a D-concentration of 60 ppm it approached the value measured at normal (150 ppm) D-concentration. In the domain above normal D-concentration (500, 1000 ppm), cell division was again inhibited. The figure also displays that the effect of D-concentration was minimal on non-tumorous myometrial cells. The above experiment, too, reinforced our previous statement that deuterium depletion has but a minimal effect on healthy cells.

Figure II. 2 demonstrates the connection between COX–2 gene expression and deuterium concentration. With Western blot analysis we measured the COX–2 content of cells treated as detailed above. Assay has shown that Dd-water (20–60 ppm) strongly inhibits the COX–2 gene expression of HT–29 human colon carcinoma cell lines. Similarly to the results of cell proliferation experiments, here too, the effect proved to depend on the concentration.

Figure II. 3 demonstrates the effect of deuterium depleted water on COX–2 gene expression in cells both basic (unstimulated) and stimulated with cytokine (IL–1), in human myometrial cell lines. As an effect of Dd-water, expression was inhibited in the case of unstimulated cells as well, but the inhibitory effect of D-depletion was especially expressed after stimulating the COX–2 gene expression with interleukine. The results confirm our earlier observations that Dd-water inhibits COX–2 gene expression.

The above results suggest that the decrease in D-concentration influences the COX–2 gene expression and the activity of PG-formation in cells. The PG-synthesis of cells is significantly inhibited by deuterium depletion: in normal (150 ppm) water the PGI_2 level was 242.9 ng/mL, whereas in 20 ppm Dd-water it was 76 ng/mL.

These findings confirm earlier results obtained during the examination of apoptosis. We suppose that increased apoptosis induced by deuterium depletion is realised *via* the inhibition of COX–2 gene expression and PG-formation.

THE EFFECT OF DEUTERIUM DEPLETION ON THE HYDROGEN ION REGULATION OF CELLS

A given molecule or element can play a regulatory role in the cell if its concentration changes in a way perceptible to the cell, triggering or halting specific intracellular processes. It may be supposed that behind the regulatory role of deuterium there lies the fact that the intracellular D/H ratio can change. Taking into consideration our knowledge of discriminative processes connected to deuterium, we supposed that the pH-regulation of cells, and processes happening through the cell membrane might be connected with the change in the intracellular D/H ratio. Thus we examined whether the decrease in D-concentration influenced these processes.

Experiments were carried out with A4 murine haemopoietic cell lines. Intracellular pH was artificially shifted into an acidic direction and we then investigated how pH is stabilized again in the media of normal and of depleted deuterium content. These experiments testify that in the media of depleted deuterium content, intracellular pH returned to the normal value much slower than cells of normal D-content. It also became clear that in this process, the amiloride sensitive Na^+/H^+ antiport system plays a key role. This observation points to the fact that enzymes and processes, well known in transducing signals and pH regulation, may be sensitive to the change in D-concentration which raises the possibility that their activity and functioning can be influenced by the change in deuterium concentration.

The biological effects of D-depletion on plant cells were investigated in the leaves of the Canadian seaweed *(Elodea canadensis)*. It was possible to follow several biological processes after having placed the leaves into a Dd-water medium.

Measuring respiration, photosynthesis, membrane potential and changes in pH both intra and extracellular unanimously pointed to the fact that in the first half hour of deuterium depletion, the plant showed such biochemical changes as if it were put in the dark. Its respiration became quicker, photosynthesis stopped and intracellular pH became alkaline, whereas extracellular pH changed into an acidic direction [12, 14]. These effects were at a maximum some 30 minutes after treatment, reactions gradually stopped later. All the above makes us conclude that plants perceive and experience it as a stress if the D-concentration of the medium is decreased. In time, cells adapt to the altered circumstances and the stress reaction gradually ceases. All this raises the intriguing possibility that in cells (both vegetable and animal) there exist such mechanisms, which are able to perceive changes in D-concentration. This induces a stress reaction of some kind which, in turn, contributes to the adaptation of cells. Taking into consideration the high degree of sensitivity experienced in tumours in a medium of decreased deuterium content, we may suppose that in tumour cells this adaptation happens slowly or not at all, which results in the necrosis of cells.

THE EFFECT OF Dd-WATER ON THE GERMINATION OF SEEDS

It is a well-known fact that following water intake, a very quick growth begins in germinating seeds. The effect of Dd-water on the germination of seeds was interesting from two aspects. The first is that this is a very simple – if not the simplest – system to examine the effect of Dd-water in, and the second is that it can provide further proof that effects observed in animal cells can be observed in plants as well.

We performed germination experiments with several plant species and about 12–14 thousand seeds, in standardised circumstances, in a medium of 20–300 ppm D-content. It can generally be stated that in a medium of depleted deuterium content, in all independent experiments, the sprout and root lengths of the seeds were less, compared to the controls. As shown in *Fig. II. 3*, we experienced a significant difference in sensitivity between the plant species (and also between breeds). While in the case of oat, the inhibitory effect of Dd-water was a mere 5 percent compared to the control, with squash this was 60 percent [12, 14, 21]. (Regarding the scatter of the experiments, a deviation greater than 10 percent from the control group can be considered as significant.)

Examining the effect of D-content on the germination of rice, we may conclude from the results that the optimal D-concentration of water for seeds falls into the natural (150 ppm) range.

It is important to stress that inhibition was at its strongest 5–6 days after the beginning of germination. Having measured sprout length after 10–12 days, the effect experienced earlier could not be observed. This confirms the results

Table II. 3 *The change in the length of plant sprouts in normal and in D-depleted water*

Plant species	Average sprout length (mm)		Δ%
	Deuterium concentration		
	20 ppm	150 ppm	
Rice	17.6	26.0	−32
Soy	35.0	44.3	−21
Wheat	44.1	51.2	−14
Sunflower	21.5	28.3	−24
Maize	30.6	40.8	−25
Barley	34.7	38.9	−10
Oat	56.0	58.8	−5
Lentil	13.7	17.5	−22
Mustard	34.5	39.9	−13
Squash	25.2	62.2	−60

of experiments carried out with heavy water, where it was stated that living organisms are able to adapt to the growing D-content of the medium within a relatively broad range. According to our results, cells are able to adapt also to

Table II. 4 *Changes in the sprout length of rice as a function of the D-content of water*

D-concentration (ppm)	Average sprout length (mm)	Δ%
20	17.6	−32
57	21.0	−19
113	23.0	−11
150	26.0	±0
193	22.6	−11
257	18.2	−29
300	17.8	−31

the D-depleted medium relatively quickly. This observation confirms our results with *in vitro* tissue culture where a more significant effect was experienced at the beginning of the treatment.

THE RESULTS COMPARED
TO THE PRESENT STATE OF SCIENCE

The practicability of a new hypothesis is defined by how it fits into the scientific results observed up till that time. With regard to this, we consider it important to analyse and interpret our results from two aspects: a) the possible role of deuterium as a mass number two isotope of hydrogen in regulation from a chemical aspect, and b) the possible connection of the change in D-concentration with the known signal transducing systems.

a) We think that the effect of deuterium from a chemical aspect has been amply studied during the past 60 years. From the vast inventory of knowledge we wish to point out those which may be of significance in justifying the role of naturally occurring deuterium in biological systems.
   - The concentration of deuterium is 12–14 mmol/L in living organisms;
   - Deuterium behaves differently from hydrogen in chemical reactions;
   - This can manifest itself in enzyme reactions as well, as for example k_H/k_D may alter between 1.5 and 10;
   - Deuterium bond is stronger than hydrogen bond;
   - This quality of deuterium is manifest in biological systems as well;
   - Great D-concentration is expressly toxic;
   - Certain enzymes and processes are able to differentiate between the two variants of hydrogen, in favour of hydrogen.

b) Irrespective of our theory, from among the results accepted in our days in connection with the regulation of cell division, we should highlight the following:
   - Prior to cell division, the Na^+/H^+ transport system becomes activated in the cell membrane, this drives H^+ out of the cell while taking up Na^+ [22]. In the course of this process intracellular H^+ ion concentration decreases (pH increases), which is a commonly observed phenomenon prior to cell division, and is said to be in a cause-effect relationship with the start of cell division. Numerous experiments confirm the conclusion that the activation of the Na^+/H^+ system is indispensable for the start of cell division.
   - Mutant cell lines were produced in which the Na^+/H^+ transport system failed to work. It was found that as a consequence of mutation, the cells lose their ability to divide at acidic and neutral pH values [23].

73

– The effect of growth factors was examined to find out through what mechanism the cell division signal gets transmitted. These experiments showed that growth factors activate the Na^+/H^+ system.
– The connection between the activated Na^+/H^+ system and the cancerous character of the cell line was proved by two series of experiments. Firstly, it was stated that in a cancerous cell line produced *via* mutation, the pH increased compared to that in the initial cell line [24].
– Further, a direct connection was found between the functioning of oncogenes and the shift in intracellular pH value, because the injection into a cell of the protein coded by the *H-ras* oncogene or the expression of the *V-mos* and *H-ras* oncogenes also shifted the pH of the cell towards an alkaline character by activating the Na^+/H^+ system [25, 26].
– Beside the Na^+/H^+ system, the activation of another H^+ transport system initiates similar changes. In these experiments the ATP-ase gene was isolated from yeast and this was used to transform a murine and a simian cell line. Gene expression and its product, ATP-ase, continuously extruded the H^+ ions from the cell, which resulted in intracellular pH increase. The really astonishing result of the above experiment was that cells transformed by the ATP-ase gene of yeast became carcinogenic [27].
– These experiments point out that to initiate cell division, the activation of a system through which H^+ ions are extruded from the cell and thus, increasing the pH is indispensable.
– It is important to note that the artificial increase of intracellular pH did not seem to be sufficient to stimulate cell proliferation [28], thus another process (e.g. the shift of the D/H ratio) must play a role in the transduction of cell division signals.

CONCLUSIONS

During experiments carried out in the past years, we have been changing the D-content of the medium in various biological systems. We experienced that in every single case the decrease of the D-concentration induced significant changes, proving that living organisms, which, through millions of years have become adapted to a D-concentration of about 150 ppm, are able to perceive the lack of deuterium.

Considering our knowledge about deuterium and the achievements of molecular biology in the past decades and also the observations of our experiments with deuterium depletion, we have set up the following hypothesis on the regulatory role of naturally occurring deuterium.

In higher living organisms, over millions of years, a regulatory system has developed which is sensitive to the intracellular changes in the D/H ratio. The

D/H ratio can increase in cells if one of the H$^+$ transport systems (H$^+$-ATP-ase, Na$^+$/H$^+$ antiport system) gets activated in the membrane. This process prefers H$^+$; thus H$^+$ and D$^+$ get ejected from the cell or organelle (e.g., mitochondria) not in proportion to their ratio of occurrence. The consequent intracellular change in the D/H ratio is "perceived" by certain enzymes, because – through the shift in D/H ratio – D$^+$ is more likely to bind to a given place in proteins. The D$^+$ bond may stabilise the conformation of the protein, rendering it either more, or less, active and this ultimately has an effect on its function. From this point on, the signal continues its way through the already known or hitherto undiscovered molecular systems and creates the well-known cell-biological phenomena, such as, e.g. cell division.

In our view, the validity of the hypothesis is reinforced by the references in literature and proved by experiments carried out with Dd-water.

On the basis of our achievements, we presume that the application of Dd-water offers a new therapeutic possibility for clinicians – complementing existing anti-cancer treatments – to treat cancer more efficiently and the preparation may play a role in prevention, too.

REFERENCES

1. *Isotope Effects in Chemical Reactions* (Collins C. J. and Bowman N. S., Eds.) Van Nostrand Rheinhold, New York, (1971).
2. Jancsó, G. and Van Hook, W. A. (1974) *Chem. Rev.* **74**, 689-750.
3. Jameson, C. J. *Isotopes in the Physical and Biomedical Sciences* Vol 2. (Buncel E. and Jones J. R., Eds.) Elsevier, Amsterdam (1991)
4. *Annals of the New York Academy of Sciences* (1960) **84**, 573-781.
5. Rundel, P. W. Ehleringer, J. R. and Nagy, K. A. *Stable Isotopes in Ecological Research.* Springer, New York (1988).
6. Katz, J. J. and Crespi, H. L. Isotope effects in biological systems. In: Collins, C. J. and Bowman, N. S. (eds.) *Isotope Effects in Chemical Reactions.* Van Nostrand Reinhold, New York (1971) 286-363.
7. Yurtsever, Y. and Gat, J. R. *Stable Isotopes Hydrology* (Gat J. R. and Gonfiantini R., Eds.) International Atomic Energy Agency, Vienna (1981) 103-142.
8. Ziegler, H., Osmond, C. B., Stichler, W., Trimborn, P. (1976) *Planta* **128**, 85-92.
9. Estep, M. F. and Hoering, T. C. (1981) *Plant Physiol.* **67**, 474-477.
10. Kotyk, A., Dvoráková, M. and Koryta, J. (1990) *FEBS Lett.* **264**, 203-205.
11. Somlyai, G., Jancsó, G., Jákli, Gy., Vass, K., Barna, B., Lakics, V., and Gaál, T. (1993) *FEBS Lett.* **317**, 1-4.
12. Somlyai G., Jancsó G., Jákli Gy., Laskay G., Galbács Z., Galbács G., Kiss A. S. and Berkényi T. (1996) *Természetgyógyászat* **10**, 29-32.
13. Berkényi T., Somlyai G., Jákli Gy., Jancsó G. (1996) *Kisállatorvoslás* **3**, 114-115.

14. Somlyai, G., Laskay, G., Berkényi, T., Galbács, X., Galbács G., Kiss, S. A., Jákli, Gy., Jancsó, G. (1997) *Erfahrungsheilkunde* **7**, 381-388.

15. Somlyai, G., Laskay, G., Berkényi, T., Jákli, Gy., Jancsó, G. (1997) In: Heys, J. R. and Melillo, D. G. (eds.) *Synthesis and Applications of Isotopically Labeled Compounds.* John Wiley and Sons Ltd. 137-141.

16. Somlyai, G., Laskay, G., Berkényi, T., Galbács, Z., Galbács G., Kiss, A. S., Jákli, Gy., Jancsó, G. (1998) *Z. Onkol/J. of Oncol.* **30**, 91-94.

17. Somlyai G. (1998) *Komplementer Medicina* **2**, 6-9.

18. Levy, G. (1997) *The FASEB J.* **11**, 234-237.

19. Smalley, W. and DuBois, R. (1997) *Advances in Pharmacology* **39**, 1-20.

20. Subbaramajah, K., Zakim, D., Weksler, B. B. and Dannenberg, A. J. (1997) *P. S. E. B. M.* **216**, 201-210.

21. Belea A., Kiss A. S., Galbács Z. (1997) *Növénytermelés* **46**, 447-485.

22. Schuldinger, S. and Rozengurt, E. (1982) *P. N. A. S.* **79**, 7778-7782.

23. Pouysségur, J., Sardet, C., Franchi, A., L'Allemain, G. and Paris, S. (1984) *P. N. A. S.* **81**, 4833-4837.

24. Sharon, S. O. and Pardee, A. B. (1987) *P. N. A. S.* **84**, 2766-2770.

25. Hagag, N., Lacal, J. C., Graber, M., Aaronson, S. and Viola, M. V. (1987) *Mol. Cell. Biol,* **7**, 1984-1988.

26. Doppler, W., Jagga, R. and Froner, B. (1987) *Gene* **54**, 147-153.

27. Perona, R. and Serano, R. (1988) *Nature* **334**, 438-440.

28. Moolenaar, W. H., DeFize, L. H. K. and DeLaat, S. W. (1986) *J. Exp. Biol.* **124**, 359-373.

CHAPTER TWO

To Doctors

A man walking in the forest meets a woodcutter who is arduously trying to saw a trunk into smaller pieces. The man goes closer to see what the woodcutter is drudging with, and asks him:
"Excuse me, but have you noticed that your saw is completely dull. Shouldn't you sharpen it?"
"I have no time for that, I have to go on sawing!"

INTRODUCTION

During my work in the past years I have experienced that most doctors are doing their best to keep up with the ever growing number of patients, and this means an ever-growing challenge for them. While for years we have been cherishing the illusion that tumour treatment was developing, the occurrence of cancer deaths has doubled since the 1960s. Patients are still subjected to grave side effects during treatment, they have to wait long for control examinations and also for radiotherapy, not everybody can afford to buy the newest drugs, and so, unfortunately, the number of cancer deaths is constantly growing. It must be very difficult to go on with the healing work when the doctor knows that his best knowledge, in most cases, is only sufficient for prolonging the patient's life or for relieving his pain.

I recommend the following chapter to doctors healing cancer patients and I ask them to take up the most important thoughts. I also ask them to stop for a moment in this big hustle and devote some time to "sharpening the tool". They should believe that the little time needed for sharpening is worth the effort. You may ask why a researcher-biologist gives advice to doctors and why he recommends an "arrogantly" simple method – deuterium depletion. The answer is simple: because we are in the possession of the knowledge of the effect of deuterium depletion and we have, for seven years been recording every observation result and experience considered to be of any significance. We spent thousands of hours listening to patients, recording their experiences, the changes in their state and tried to find out the possible connections between various factors.

Why have we done all these? The answer is: to be able to hand over the collected experiences for everyone's benefit.

I do not think that everything written down here is final and unchangeable from every aspect. Numerous questions are still to be answered. **This summary was intended as a starting point for a later extensive application of Dd-water.**

With time the majority of the observations and the main principles of application will be confirmed, although there might be some fields where we will have to change the course described here.

I am convinced that the scope of doctors' duties will alter in the new Millennium. They will not have to use every effort to make maximum use of the available therapies while the patient's condition worsens but, instead, give them every help on the way to recovery. I do hope that within a short time the number of successful treatments will multiply.

The history so far of deuterium depletion can be divided into three main phases:

"Ancient (or Golden) Age" (1992–1993): the first patients. In those days we could produce "greater" amounts (30 L/day) of 110–115 ppm Dd-water – in contrast to the 150 ppm D-content of natural waters. Lacking experience, there were significant differences between individual patients, there were no dosage aspects, and we could only reach a minimal decrease of D-content in the patients' bodies. Despite all these, it was clear as soon as the first patients were involved that the consumption of deuterium depleted water had a significant and positive effect.

The Middle Ages (1994–1998): Then it was possible to produce 90–95 ppm Dd-water in great quantities (4500L/day). It was in this period that *Phase II* clinical trials began, and the adjuvant treatment of a greater number of patients was possible. Due to the lower D-concentration, the effectiveness of Dd-water improved considerably. By increasing the daily dose of Dd-water, we could ensure a continuous decrease of D-concentration in the body for months. In the meantime, the constraints of the treatment also became clear. In patients with a considerable tumour mass (e.g. tumour of the colon with metastases of 5–6 cm in diameter in the liver), a considerable improvement or stagnation following Dd-water consumption was experienced, in some cases for up to 16–18 months. The D-concentration applied, however, was not sufficient for destroying all cancer cells. The cells, namely, which had for months been in a medium of lower D-content (and because no further decrease in D-concentration happened), slowly got adapted to the new medium and divided further. Then the patients again showed signs of progression. The effectiveness of the treatment considerably improved when – although in smaller quantities – Dd-water of 45 ppm was at our disposal. Then we tried to improve the effect by joint application of two kinds of water with different D-contents.

The New Era. The third period began in the summer of 1998 when a 20–30 ppm Dd-water became available. Applying various mixing rates of Dd-water and normal water, this ensured a longer period of decreasing D-concentration in patients. The application of the substance further improved our results.

78

The majority of observations described below are built on the application of 90–95 ppm Dd-water in the first place, but contain the experience gained since the summer of 1998 with 20–30 ppm Dd-water as well.

I am convinced that the knowledge and information at our disposal, which we would like to share, can improve or even save the lives of tens of thousands. More than seven years have given us the opportunity to record observations and to draw conclusions. It is a great pleasure through this book to be able to convey our knowledge to medical practitioners.

HOW TO DEFINE THE EFFECTIVENESS
OF A POTENTIAL DRUG

We do not wish to go into details about the complex system of drug development, but would like to highlight one aspect only. If a drug for registration passes the first trials followed by further ones confirming its effect, and it does not fail in the toxicological tests either, then, after the work of 6–8 years, the moment arrives when it has to be proved, in clinical tests, too, that the preparation induces the effects supposed on the basis of pre-clinical results, in humans, too. For this a strict system of requirements has been set up during the past decades. The aim of the painstaking preparatory process is to be able to state with absolute certainty that the changes experienced by a patient following the intake of a preparation are in fact in a cause-effect relationship. If this convincingly follows from results, and toxicological tests also allow the human consumption of the product (with anti-cancer preparations this does not mean that the potential drug is non-toxic), the product will be approved and the preparation becomes a drug.

In the case of anti-cancer drugs, clinical trials are especially important. One reason for this is, that most preparations are highly toxic, their application endangers the patient's life and in some cases can even lead to the appearance of further tumours. With these negative side-effects, only the follow-up of several hundreds or thousands of patients for years can provide the necessary information with which it can be stated that the preparation is more effective than the existing ones, or that the same effect can be reached by using it but with less side-effects. The development of anti-tumour drugs has been proceeding year by year, with thousands of clinical trials, testing several thousands of future drugs, improving survival rates percent by percent. Successful clinical trials meant in most cases that, involving several thousand patients, it was proved that the given drug combination improved survival rates by 5 percent. This means that while the tumour allowed an average survival of 9 months, due to the new drug these patients died 50 days later, on the average. The present system and strategy of drug development is further complicated by the fact that a great percentage of patients reacted very well to a certain future drug that manifested itself in quick tumour regression, the average survival rate, however, did not improve at all.

The development of anti-tumour drugs and the concomitant system of clinical trials is similar to the construction of machines unfit for flying and to the complex system of tests to examine which of them flies further. Neither can be used for flying because they were not constructed with a knowledge of the basic laws of air current, but depending on the formation and the choice of testing, certain differences can be found between various constructions. If, for instance, the construction is light (the drug has no serious side-effects), it can fly further or, if released from a high mountain (treatment with a greater dose) it can also "fly" further. One thing is sure, the place of landing is rather uncertain, and in any case it is very near to the place of takeoff, whereas "landing" bears in itself the possibility of serious injuries or maybe the total destruction of the construction.

If somebody has been testing constructions unsuitable for flying throughout his entire life, following the old schemes, and suddenly beholds a construction that flies much farther than the other ones, he is ready to think that there must have been a methodological mistake somewhere. He does not rejoice to see that at last, there is something that leaves the others behind, but he begins to look for mistakes in the testing methods. The original aim, namely, to create a flying construction, becomes degraded to a secondary one. After a series of failures there remains the "excellent" and sufficiently complicated testing systems. And here come the excuses: the construction did not fly farther than the others on enough occasions (the small number of patients); certain navigation circumstances were not identical with those of the other group (the group of patients is heterogeneous); and on very rare occasions the old machines also succeeded in flying that far (conventional therapy can also bring good results), etc.

We have been listening to these criticisms for the ninth year now, while we have repeatedly proved, according to various testing aspects, that deuterium depletion is an effective means for curing cancer patients. This chapter can be understood and accepted only if the reader breaks with the tendency of development that has proved to be a dead end, and tries to accept that the anti-cancer medication based on deuterium depletion works on a principle with the help of which one can "fly". To be able to state this, one does not need a statistical analysis, double blind clinical trials, several thousand patients or a follow up period of 3 to 5 years. The effect of the future drug based on Dd–water can be seen immediately (in 1–2 months) after the beginning of the treatment. After so many failures and fiascos, in the end we are able to build constructions suitable for flying – there must be a time when the first "real" anti-cancer drug will be produced.

80

THE THEORETICAL BACKGROUND OF DOSAGE

The method of the application of Dd-water and mainly of its dosage is basically different from that of cytostatic treatment schemes used today. It is generally known that the dosage of cytostatic drugs is established on the basis of their proportion to body weight or body surface, and this same amount is applied throughout the entire protocol. There can, of course, be some alterations depending on the patient's reactions, but the generally accepted principle is to run the whole treatment with a constant amount of the cytostatic drug.

Research so far has proved that the anti-cancer effect of Dd-water resulted from the D-concentration downshift that is brought about when an amount of water with a higher D-concentration (the water in the body of the patient) is mixed with a certain amount of water with a lower D-concentration (Dd-water-based preparation). *In vitro* experiments showed that the cells, after a certain time (6–10–20 hours), were adapted to the new D-concentration, which was lower than before. The adaptation means that the inhibition of cell division ceases to exist. Through *in vitro* experiments it has also been proved that in cases when the lower level of D-concentration was brought about step by step, the inhibitory effect was twice as strong at the level of DNA synthesis; and also that minimally three times longer inhibition could be established. Numerically this means that diminishing D-concentration in one step (from 150 ppm to 30 ppm) resulted in 15–20 percent inhibition of the incorporation of H^3-thymidin, and the effect gradually ceased to exist within 16–20 hours. If, however, the diminishing of D-concentration took place in 5 steps from 150 ppm to 42 ppm (150ppm–60 ppm–55 ppm–46 ppm–42 ppm), within 48 to 72 hours, the inhibitory effect became stronger day by day and reached 40 percent.

When establishing the dosage, we have to keep in mind that the main task is to provide an approximately similar daily level of decreasing deuterium concentration for a longer period, i.e., for weeks or months, and, if this becomes impossible, to maintain a low level of D-concentration. In practice, the highest level of diminishing D-concentration in the patient's body can be reached just after the patient has started drinking Dd-water. With a decreasing D-concentration established in the patient's body, later, the same amount of the daily dose causes a relatively smaller change in the level of diminishing D-concentration. In order to maintain the necessary daily decrease of D-concentration, after a certain time (1–2 months), treatment should be continued with Dd-water of a 6–10–15 ppm lower D-content. Then the previous concentration gradient is restored between the patient's body and the Dd-water to be consumed, which results in a further decrease of D-concentration. In an ideal case, the D-concentration of Dd-water should be fixed daily, but practice has shown that a change in every 1–2 weeks or 1–2 months is sufficient. Thus, the concentration gradient is maintainable despite the fact that the D-concentration of the patient's body had dropped significantly.

In the beginning of the treatment, when establishing the dosage, the main factor is body weight. The amount of daily dosage will change during the months of treatment, because our aim is to establish a constant change of D-concentration with Dd-water, to reach a necessary level for the optimal effect but at the same time the decrease in D-concentration should be kept for the longest possible time. If this is no longer possible, our further aim is to keep the D-concentration in the patient continuously at the lowest possible level.

In addition to determining the minimal dose for the desired effect, there is a further important aspect to be kept in mind during treatment. The effect of Dd-water is quite direct and will be apparent in a short time, in most cases the reaction is very quick and explicit. The rapid manifestation is the necrosis of the tumour, which requires the body to give a complicated answer, a solution of many steps. An inflammatory process starts in the necrotising tumour, which can be a very difficult task for the patient's body to cope with. The structures of the tissues must undergo a transformation, some pieces of tissue can detach if the location of the tumour makes it possible (e.g., in the colon or in the bladder), and it will be necessary for that part of the body to be reconstructed. One of the most important rules in the application of Dd-water is that the process of recovery must not be hurried. The body must keep in step with "cleaning up" of the necrotising tumour mass. The details of this process will be discussed later.

DOSAGE

"Dose" (here too, as with other drugs) has a double meaning: it refers to the daily allowance of Dd-water but it also defines its D-concentration (the active agent content of the preparation).

It is desirable that the patient drink at least 75–80 percent of his daily water intake in the form of Dd-water. An occasional broth made with normal water or a limited consumption of fruit and vegetables is allowed but we do not recommend the consumption of tea, milk, wine, beer, fruit juices or other liquid in great quantities (tea can be prepared with Dd-water as well). We also recommend that after consuming food of a normal D-content, the patients drink Dd-water to keep the mean D-concentration of meals below the normal level.

It is easy to see that just like with other medications, the dose can be raised in two ways: by increasing the daily amount of Dd-water or by lowering its D-concentration (i.e. increasing the active factor content).

Dosage must be determined individually, according to the patient's condition. In general it can be stated that there should be a minimal, 50–60 ppm difference between the D-concentration of the patient's body and that of the Dd-water to be consumed.

The "prescription", thus, should contain two numbers, the first referring to the volume (L), and the second to the D-concentration of the preparation (ppm).

EXPERIENCES WITH 90–95 PPM Dd-WATER – THEORETICAL CALCULATIONS

Until the application of 25-ppm Dd-water, the dosage was based on experiments carried out with dogs and cats until the summer of 1998, as well as on the effective dosage of Dd-water used during the *Phase II* clinical trial and adjuvant therapy. With these experiments we arrived at a range of 0.014–0.026 kg Dd-water/body mass kg/day, applying Dd-water with a concentration of 90–95 ppm. Considering this wide range, a further division was necessary. *Table II.5* shows that by taking two extremes and one middle value, the starting dose of Dd-water was established as a function of the body mass.

Taking the above doses into consideration, we will demonstrate with some examples the theoretically established D-concentration in the human body after the consumption of Dd-water. In the calculations we assume that the water content of the human body is approximately 60 percent; the D-concentration applied is 95 ppm; and the D-concentration of the water content of the body at the beginning of treatment is 150 ppm.

Taking into account our results when we determined the D-concentration of the serum of a dog having consumed Dd-water as drinking water, or the D-concentration in the urine of the patients, we can say that approximately 50 percent of the theoretically calculated D-concentration could be realised. This means that the results of the theoretical calculations concerning the changes in the daily D-concentration values, will be realised only in 50 percent. This is also true for the diminishing total D-concentration due to the long-range Dd-water consumption.

From the examples we can see that the three dosage schemes differ from one another especially at the beginning of the treatment. As the treatment goes on, there will be an alteration in the obtainable absolute reduction of the D-concentration; nevertheless, the change in the daily D-concentration is almost the same in the three cases.

As mentioned before, one of the aims of dosage is to keep the daily D-concentration change at a constant level. Here you can find an example of how an increase in the interim dose can influence changes in the daily D-concentration: in *Table II. 6* it can be observed that the change in D-concentration after 29–30 days of Dd-water consumption is only 0.45 ppm if the patient consumes the Dd-water according to the DDW-A scheme. However, if we put the patient on the 30th day from the DDW-A scheme to the DDW-B, and from this time on he will drink 1 litre of Dd-water daily, theoretically there will be a 1.1 ppm of D-concentration decrease on the first day, and this is almost equal to the reduction of D-concentration established at the beginning of the treatment.

Table II. 5 *Dose regimen with 90–95 ppm Dd-water as a function of body mass and three different doses related to 1 kg of body mass*

Body mass (kg)	D-content: 90-95 ppm		
	DDW-A 0.014 kg DDW/body mass kg/day	**DDW-B** 0.020 kg DDW/body mass kg/day	**DDW-C** 0.026 kg DDW/body mass kg/day
10	0.14	0.20	0.26
20	0.28	0.40	0.52
30	0.42	0.60	0.78
40	0.56	0.80	1.04
50	0.70	1.00	1.30
60	0.84	1.20	1.56
70	0.98	1.40	1.82
80	1.12	1.60	2.08
90	1.26	1.80	2.34
100	1.40	2.00	2.60

Table II. 6 *Change in D-concentration (in ppm) on days 0–1, 29–30 and 99–100 of treatment*

Days	DDW-A		DDW-B		DDW-C	
	D-conc.	**ΔD-conc.**	**D-conc.**	**ΔD-conc.**	**D-conc.**	**ΔD-conc.**
0	150		150		150	
1	148.74	1.26	148.22	1.78	147.7	2.3
29	127.8		122.96		119.37	
30	127.35	0.45	122.5	0.46	118.91	0.46
99	111.6		107.79		105.39	
100	111.5	0.1	107.69	0.1	105.3	0.09

CALCULATING THE DOSE WHEN APPLYING 25-PPM Dd-WATER

In the "New Era", the application of Dd-water of a 25-ppm D-content enabled us to establish a more effective dosage regimen. We diluted Dd-water adding normal water to it, depending on the state of the patient, the tumour type, the duration of the treatment, etc. Our recommendations as to what D-content the applied water should have will be dealt with later. In *Table II. 7* the D-content of the water is given if it is diluted with various mixing rates.

The final concentration of a given mixture, knowing the D-concentration of its two components, can be calculated with the help of the formula below:

$$V_1 \times C_1 + V_2 \times C_2 = (V_1 + V_2) \times C_v$$

where

C_1 = the D-concentration of one component in ppm
C_2 = the D-concentration of the other component in ppm
V_1 = the amount of one component in litres
V_2 = the amount of the other component in litres
C_v = the final concentration of the mixture in ppm

Since C_1, C_2, V_1, and V_2 are known quantities, after performing the operation we get the value of C_v, i.e. the final concentration of the mixture.

By applying the above formula we may also define how the D-concentration in the patient's body changes subsequent to consuming Dd-water.

Example: If a patient with a body mass of 70 kg, on the first day of the cure drinks (V_2) 1.4 litres of 87.5-ppm Dd-water (C_2 = 87.5 ppm, i.e. 25-ppm water mixed with normal water in a 1:1 ratio), the theoretical decrease of the D-concentration can be calculated as follows:

We suppose that 60 percent of the patient's body is made up of water the D-concentration of which (C_1) is 150 ppm, thus 1.4 litres of Dd-water (V_2) are diluted by 42 litres (V_1 = 0.6 x 70 kg) of normal water.

$$42 \times 150 + 1.4 \times 87.5 = (42 + 1.4) \times C_v$$
$$6300 + 122.5 = 43.4 \times C_v$$
$$6422.5 = 43.5 \times C_v$$
$$6422.5 : 43.5 = C_v$$
$$147.64 = C_v$$

This calculation shows that by the end of the first day, the D-concentration of the water in the patient's body will be theoretically 2.4 ppm lower. As stated previously, it is only about 50 percent of the theoretical value that can be realised. In our case this means that by the end of the first day, D-concentration will be approx. 1.2 ppm lower.

Table II. 7 *The D-concentration of Dd-water gained by mixing 25-ppm*
Dd-water and 150-ppm (normal) water

25-ppm (Dd-water)	150-ppm (normal water)	
Mixing rate		Final concentration (ppm)
1	9	137.5
2	8	125.0
3	7	112.5
4	6	100.0
5	5	87.5
6	4	75.0
7	3	62.5
8	2	50.0
9	1	37.5
10	0	25.0

FACTORS INFLUENCING THE DETERMINATION OF DOSAGE

1. The volume of Dd-water consumed. Based on many years of experience, it became obvious that there is a clear-cut connection between the volume of the Dd-water consumed and its efficacy. The more Dd-water is consumed with the less D-concentration, the greater the decrease of D-concentration in the patient's body, thus the stronger the effect. It is very important that the dose should be established accurately, so that the patient drinks enough Dd-water, to get the necessary effect. At the same time, it is important to keep the decreased D-concentration in the body as long as possible.

2. The concentration of Dd-water. The lower the D-concentration of the applied water, the greater the effect from the consumption of one unit of Dd-water. However, it does not follow from this that the patient should be given water of a low (30 ppm) D-concentration right at the beginning of the treatment.

3. The reaction of the patient to the administration of Dd-water. Based on experience gained during the last years, it can be stated that Dd-water is not toxic, it does not cause any destruction in the blood count or in the mucous membrane, it does not cause nausea, etc. If Dd-water is consumed by a healthy person or by a patient in total remission, there are no observable side-effects of consumption. But there can be some subjective and objective side-effects of drinking Dd-water by a tumour patient. These effects are caused by the interaction of Dd-water and the tumour. Therefore it is the physician's task to follow both direct and side-effects of Dd-water. One of the most typical side-effects is the appearance of drowsiness and fatigue within one or two weeks or in some cases later. The drowsiness and fatigue disappear after some weeks, depending on other parameters too. When the dosage is established, it is most important to consider the strength of the reaction and should it be too strong, the daily dose must be reduced by 10–20 percent. It is not advisable to interrupt the consumption of Dd-water, because this will make it easier for the tumour to get used to the Dd-water medium. If the patient no longer feels side-effects or feels that the effects have become weaker, then the dose must be increased to the original level.

Naturally, the opposite of the above may also happen, namely when there is no effect at all. If the application of Dd-water has no perceivable effect within a month, a more significant (40–50 percent) increase of the dose should be implemented, which means either the increase of the volume of Dd-water or the administration of a preparation with a D-content lower by 10–15 ppm.

It is important to stress that hypersomnia or fatigue appears mostly in an advanced condition and in the presence of a relatively large tumour (3–6 cm in diameter). If the tumour is of a smaller size, the above symptoms may not appear at all.

4. **The body mass of the patient.** Children are the most ideal subjects, because a relatively large dose can be administered on account of their relatively small body mass. If the application of 85–90 ppm Dd-water is considered, we can say that up to an average weight of 60–70 kg, the daily fluid requirement and the recommended daily dose of Dd-water are in accordance with each other. It is more difficult to remain within the limits of the daily fluid need of patients within the 80–100 kg range, so, for them it is by all means necessary to drink Dd-water of a lower D-concentration so that the established dose can be kept for a longer period of time.

5. **Types of tumours and their sensitivity to deuterium depletion.** From our experience and according to expectations, it can unanimously be stated that tumours of different origin or those belonging to different classes do not have the same degree of sensitivity to deuterium depletion. Based on our data we established a scale of sensitivity, which, of course, may change with the information we will obtain from more patients with different types of tumours. The types of tumours in relation to their sensitivity can be seen in *Table II. 8.* Additional criteria in connection with the sensitivity of the tumours are defined below, from among which one or more may be characteristic of a given tumour type.

Types with high sensitivity
- Dd-water proved to be effective in 70–80 percent of patients in quite a lot of patients (30–60).
- In some cases, tumour regression became evident within a few weeks.
- We succeeded in stopping or reversing progression of the disease even in its most hopeless state.
- The effect was quick and obvious with some classes of tumours, where we had data of a few patients only (10–50).
- Dd-water did not only prolong the life of the patients, but it resulted in obvious tumour regression and remission.

Types with average sensitivity
- In cases where the volume of the tumour was not significant (less than 3–4 cm in diameter), the effect of Dd-water was proved without doubt by the regression of the tumour.
- On the effect of Dd-water consumption, the progression of the disease was stopped with patients in the late stage of the illness.

Types with less sensitivity
- From among several tumours of the same type only some patients showed a positive reaction.
- Regression did not occur; we only succeeded in stopping tumour growth.

Resistant tumours
- In these types there are no cases where the efficacy of deuterium depletion can be confirmed (we note however, that these types of tumours have been encountered only rarely).

We had some patients with other tumour types, too; so far we are unable to make any statement concerning the sensitivity of these types. They are considered to be more or less sensitive to Dd-water, but the small number of cases makes it impossible to establish an exact classification; such are tumours of the hard palate, pharynx, oesophagus, bone, various types of sarcomas of the soft tissue, neuroblastoma, etc.

Table II. 8 *Sensitivity of different tumour types to deuterium depletion*

Types with high sensitivity	Types with average sensitivity	Types with lesser sensitivity	Resistant types
uterus	colon	gallbladder*	pancreas*
cervix	rectum	melanoma	
tongue*	lungs	glioblastoma	
larynx*	myeloma multiplex		
thyroid glands*	astrocytoma		
leukaemia	Hodgkin-lymphoma		
ALL	NH-lymphoma		
AML	ovaries		
CLL	bladder*		
CML	liver*		
breast			
skin			
testis*			
stomach**			
prostate			
kidneys*			

* Experience based on relatively few cases (2–10)
** The higher degree of sensitivity may be due to the direct contact of the tumour with Dd-water.

6. Total mass of the tumour. Here too, the relation is obvious: the smaller the mass of the tumour, the higher the efficacy of Dd-water. During the establishment of the dosage, a small mass of tumour means that a lower dose may be enough to reach regression, and at the same time the necrosis of a smaller mass of tumour does not burden the body so much. Thus a dose larger than necessary can be applied at the beginning of the treatment. In case of an advanced stage, while establishing the dosage, besides the aim of keeping the efficacy, the long duration of Dd-water consumption must also be kept in mind.

7. The shape of the tumour. Our experience indicates that the part of the tumour which is in contact with the healthy tissues is the most sensitive to Dd-water. This may be explained by the fact that this is the most invasive area of the tumour, where the cells are in the most sensitive stage of cell division to deuterium depletion. Our experiences have shown that during the regression of the tumour, the tentacles infiltrating the surrounding healthy tissues are retracted first. This may have an importance especially when, e.g. the patient can be pre-treated with Dd-water before surgery. This pre-treatment may make the removal of the tumour easier and the patient more operable. Drinking Dd-water can especially be advised 2–4 months before the operation of brain tumours, if otherwise there are no contraindications.

It is also characteristic that the tumours of a bigger expansion, e.g. the tumours of the pleural membrane, are much more responsive than those of a globular shape.

8. The localisation of the tumour. From the point of view of the effect, the most "ideal" tumours are those which can get in direct contact with Dd-water. These are tumours of the stomach and the oral cavity, or tumours and metastases of tumours close to the skin surface. This is why 50–60 percent of the dosage based on body mass can be enough to be effective in the beginning of treating a patient with stomach cancer. In the case of tumours of the oral cavity we suggest keeping the Dd-water in the mouth for a while (5–10 minutes, if possible) before swallowing it, 5–7 times a day. We suggest applying Dd-water based ointments or wet pack saturated with Dd-water for the treatment of tumours located close to the body surface. We must take into consideration that while wet packing, for example, there is a significant decrease in D-concentration locally, but after finishing the cure, D-concentration will increase quickly, thus establishing favourable conditions for the tumour. This is why we suggest that one treatment should only last for a shorter time (10–15 minutes) but it should be repeated 6–8 times a day. Local treatment should coincide with the consumption of Dd-water.

9. Sensitivity of the tumour to deuterium depletion. While establishing the dose, we must take into consideration that for the treatment of more responsive tumours, in the beginning it is enough for the patient to drink Dd-water of about 90 ppm concentration, whereas patients with less responsive tumours will have to be dosed initially with Dd-water of a lower D-content. With resistant tumours and those of a lesser sensitivity, our suggestion is the consumption of 55–60 ppm Dd-water. Cancer of the pancreas, late stage melanoma, and various types of sarcoma belong to this category.

10. Treatment of primary tumours and/or metastases. It can generally be stated that primary tumours are responsive to Dd-water and their metastases are sensitive as well. When establishing the dosage, we can base the dosage on the mass of the tumour. This mass is significant if metastases have already appeared. It is obvious that the genetic heterogeneity of the tumour is also decisive, as it may lead to a resistance to deuterium depletion. Significant differences in behaviour were found between the primary tumour and its metastases.

90

According to our observations, the liver metastasis of breast tumour is more responsive to Dd-water treatment than the primary tumour itself. The metastasis in the lungs can also be sensitive but here a longer time is needed to obtain a favourable effect (8–10 months), while in the liver total regression occurred within 1–2 months.

We have to mention that the primary tumour of the gallbladder was, in two cases, responsive to treatment, while the metastases in the liver remained resistant.

11. The stage of illness of the patient at the beginning of the treatment. The earlier the Dd-water was applied, the better the chances the patient had and the more successful the results. Thus, we can draw the conclusion that it would show very good results if patients after a successful operation or chemotherapy could receive Dd-water as post-operative treatment, because when the number of still existent tumour cells is the smallest in the body, Dd-water treatment could significantly decrease the appearance of new tumours (recidiva/ metastases).

12. The patient's general physical state. If the patient's physical state is satisfactory, it does not affect dosage. With a patient, however, whose physical state is deteriorated, a subjective further deterioration may occur: after starting the treatment with Dd-water, the patient becomes more prostrate, depressed and fatigable. This does not mean a general progression of the illness, but it is a consequence of the necrotising tumour. This period can last from 1–2 weeks to some months. It is advisable to discuss the possible consequences of the treatment with the patient in advance.

13. Other treatments. The majority of patients involved in the treatment with Dd-water received conventional treatment as well. There were some exceptions, with patients who had used all the possibilities of conventional treatment, or had never started any. Generally, Dd-water did not interfere with any conventional treatment. This, however, may not be true inversely, because the side-effects of conventional treatment have several times impaired the effect of Dd-water. In the present state of the registration procedure, however, it must be accepted that the two treatments go together. Chemotherapy may especially deteriorate the effect of Dd-water treatment, if, because of side-effects, the patient is unable to consume liquid for several days. However, in several cases, patients consuming Dd-water tolerated cytostatic treatment better. Their blood count had not deteriorated and other side-effects did not occur as much as expected. Therefore, the establishment of the dosage should not be influenced by the applied cytostatic and radiotherapeutic procedure. Following surgery, the dose of Dd-water must be decreased by 60–70 percent for 6–8 days; we have observed that in some cases Dd-water slowed down the process of wound healing. We note, however, that in many cases we did not experience the above. Decreasing the dose for some days after the operation is thus indicated as a precaution.

14. The blood count of the patient. Direct examinations concerning the relation of the patient's immunological state and the drinking of Dd-water have not been performed. But on the basis of general experience, the recovery rate is better with patients having a normal white blood cell count than with those having a deteriorated blood count caused by cytostatic treatment. One of our patients with lung cancer had been treated conventionally for several months and as a result of this, the illness came to a state of stagnation. Regression was seen only months after the cytostatic treatment had been finished.

The drinking of Dd-water did not essentially influence blood counts of the patients, but with patients receiving chemotherapy, blood counts became even better than expected.

15. Time passing since the beginning of the application of Dd-water. At a more advanced stage of treatment, the amount of Dd-water should be increased and/or the administering water of a lower D-concentration should start. The more time passes from the beginning of the cure, the lower the D-concentration in the body. Time passing also means that part of the cells inside the tumour have spent a longer time in a medium of lower D-concentration, and were enabled to adapt slowly. If a certain time has already passed (this can be from some months to 1–1.5 years) and it is not possible to obtain any more decrease in D-concentration, the tumour may make further progression. This is the case of patients whose total regression could not be achieved with Dd-water.

The opposite of this, however, can also happen, when D-concentration in the patient's body can no longer be decreased, there still occurs a significant improvement. The explanation for this might be that D-concentration kept much lower than natural also gives improvement, as tumorous cells "get exhausted" in this medium.

Based on our experiences, patients who have become free of the tumour after a treatment of Dd-water can feel safer if they consume Dd-water for an additional year. Of course, the longer the duration of the remission, the more we can hope to achieve complete healing. At the first attempt to stop or interrupt the cure, we suggest doing it gradually, the duration of the deprivation should be 1.5–2 months. It is definitely suggested to repeat the cure after a pause of 2–3 months. In case of a repeated protocol, first we suggest a quick decrease in D-concentration for 4–6 months, followed by a slow cessation. Then a 4–5 months break again is recommended, followed by another 2–3 months' treatment, which can also be repeated.

FURTHER ADVICE AND RECOMMENDATIONS ON THE APPLICATION OF Dd-WATER

1. Increasing the dose. From the point of view of treatment, it is decisive how susceptible the tumour is to deuterium depletion and how quickly it can adapt to the decreasing D-concentration. Observations gained during the last years enable us to provide certain essential facts concerning this field.

Supposing we were to grade various tumour types, melanoma seems to be the quickest to adapt. With melanoma patients having a growing number of metastases, there is no time for changing the initial dose after 3–4 weeks only. We, therefore, recommend to increase the dose at the beginning of every 1–1.5 weeks, or to decrease the concentration of Dd-water given to the patient gradually. This may happen to be sufficient. Since we have 25-ppm Dd-water at our disposal, it has not been possible to test this, so we cannot tell whether with melanoma patients by increasing the dose in the above way it is sufficient to stop progression. It is also important that treatment should begin with Dd-water of a concentration below 85 ppm.

Cancer of the pancreas also belongs to the resistant category. This is the only type of tumour where, in a few cases only, we did not experience any positive change when applying Dd-water of a 90–95 ppm concentration.

With cancer of the prostate we have experienced that the tumorous cell population gets heterogeneous rather quickly, and bone metastases also appear relatively soon. Thus, in tumours of this type, we recommend a dose higher than usual and a more frequent increase, especially if the volume of bone metastases is considerable.

The opposite of this could be observed with e.g. cancer of the root of the tongue. Patients reacted very well with a given dose and the effect could be maintained for years, and the patient continued to remain in remission. When the signs of progression appeared again, the tumour again reacted to the increased dose and a repeated regression occurred. With two patients this was observable even after 4–5 years, suggesting that the genetic variability of such tumours is minimal, the cells less sensitive to deuterium depletion do not get selected in the course of years.

In general it can be recommended that the D-concentration of Dd-water should be decreased by 10–15 ppm every 1–2 months.

2. Long-term positive effect of deuterium depletion. On the basis of the long-term follow-up of patients it can be stated that neither acute nor chronic side-effects occurred during treatment. We have made two interesting observations with respect to the tumour.

The statement below seems to contradict the above, but in fact we would like to record it: in some cases it happened that after 3–4 years the patients interrupted the consumption of Dd-water despite the fact that the tumour had still been present in their bodies. These cases proved that if D-concentration was kept low enough for a considerable amount of time in the patient's body, a resistance for months was maintained even if in the meantime D-concentration again increased to the normal level. This observation at the same time is in accordance with an earlier experience, namely, that Dd-water may be effective even if the amount of deuterium does no longer decrease in the body, but remains at a low level for a longer time. In this case we think that tumour cells which had not been destroyed, adapted to the low D-concentration, and although they do not necrotise, low D-concentration nevertheless inhibits uncontrolled cell division. After some years, the slow increase of D-concentration supposedly induces stress similar to what the decrease of D-concentration did at the beginning of the treatment.

We have to stress that the above could not be observed if the patient had stopped the treatment after consumption for some months only.

3. The application of strongly deuterium-depleted water (25–50 ppm) is not advised at the beginning of the treatment. Because of misunderstanding of the prescriptions, two patients began to drink undiluted, 25-ppm Dd-water. In these cases we experienced an inflammatory reaction within the whole tumour, but despite this, no significant tumour regression happened during the period of one month. Based on the above, we suggest that the D-concentration applied should not be lower than the level needed to gain the desired effect, as later, during treatment it is possible to further decrease the D-content.

4. How long should Dd-water be consumed? The main aspect is that the treatment should have no risks, but with several patients a progression occurred within 2–3 months with the untimely cessation of the treatment. If the patient is considered as cured from a medical point of view, it is advisable to continue the cure for a minimal 1–2 years, as this means no risk whatever. One of our main tasks in the near future is to define, through appropriate experiments, the post-treatment strategy by which the patient can, with great certainty, be called cured.

5. When increasing the dose, prostration and fatigue may re-appear. If the treatment is continued with water of a lower D-content, phenomena experienced at the beginning of the treatment, i.e. hypersomnia and fatigue, may re-appear. This, however, should be considered as a positive sign, as it means that the further decrease of D-concentration induces a renewed reaction, causing the further regression of the tumour.

6. Pre-treatment with Dd-water. As mentioned before, and stressed here again, after considering the individual case, and if medical viewpoints do not contraindicate, a Dd-water treatment 4–6–8 weeks prior to surgery is suggested. Such pre-treatment of the patient may, to a great extent, contribute to the success of the operation. As a consequence of the treatment, it is not only the size of the tumour that may change but its shape may also become similar to that of a benign tumour. In our view, this may be of particular significance in the case of brain surgery where the size of the part removed is critical.

7. Daily dosage of Dd-water. It is important that the patient should cover the bulk (75–80 percent) of his daily fluid need by Dd-water. Our experiences show that it is favourable to consume the daily dose divided into portions throughout the day at equal intervals, rather than to drink it all at a specific time of the day. We also recommend that the patient drink 0.5–1 dl of undiluted Dd-water of a low D-content after consuming food of a normal D-content.

8. Pain relief. It is a known fact that in case of tumorous diseases pain relief and a better quality of life are serious challenges. A significant pain relief may accompany the consumption of Dd-water, which, in the first place, is due to the positive changes in the basic condition. It is important to note, however, that at the beginning of the treatment with Dd-water, the measure of pain may vary significantly. This transitory period may, in some cases, last for 2–3 months.

9. The diet of the patient. The long-term follow-up of several hundred cases enabled us to draw some conclusions on the correlation between consuming Dd-water and the eating habits of the patients. Our most important conclusion is that a strict vegetarian diet will decrease or totally suppress the effect of Dd-water. Neither the Gerson diet nor other kinds of drinking protocol can be suggested during Dd-water treatment. Our general suggestion is that the patient should eat normally, like a healthy person. We therefore suggest a well-balanced diet, free from extremes. We suggest the exclusion of pork and beef, but it is advisable to eat poultry, fish, and ample portions of vegetables and fruit.

Concerning eating exotic fruits, we have to mention that the deuterium content of these is much higher than that of fruits cultivated in our climate. This is why we suggest that the patient eat smaller amounts of these kinds of fruit, because they may reduce the efficacy of the treatment. We thus suggest the consumption of fruit and vegetables grown in Europe.

10. Other additional therapies. Besides conventional treatment and the consumption of Dd-water, several patients have tried other, alternative treatments to increase the chance of healing. The effect of these alternative treatments seemed evident in certain cases (influencing either positively or negatively the effect of Dd-water), but in some cases we could only suppose some mutual effects. We have collected our opinion in two groups:

Issues impairing the effect:
- Gerson diet;
- Strict vegetarian diet;
- Cure of drinking fluids;
- Consumption of coenzyme Q_{10};

– Hot sitting baths;
– Depressed state.

Issues improving the effect:
- Well-balanced mixed diet;
- Cold bath (1 minute sitting in the tub or cold-hot alternating shower);
- Local cooling of the tumour;
- Positive way of thinking;
- Strengthening of the immune system.

We do not wish to take a stand on methods or procedures applied by the patient. According to our general opinion, methods with an effect of blocking the cell cycle at a given point may make tumour cells resistant to Dd-water, as the cell is not able to enter into the phase sensitive to deuterium depletion.

11. A break in the treatment. It has happened that because of technical reasons, carelessness or change of mind, the patient stopped drinking Dd-water for a shorter or longer period. In such cases the signs of progression soon became evident. It can generally be stated that the process is reversible. If the effect of the treatment had already appeared, the treatment that continued after a break will also have results. According to our experiences, though, it is more difficult to ensure the effect if the treatment was stopped and started again several times. In our opinion, one of the most important rules of applying Dd-water is that the cure must not be stopped.

RESULTS OF HUMAN TRIALS (1992–1999)

Results of human trials come from two sources:

1. Double-blind *Phase II* clinical trial
2. D-depletion on the basis of compassionate use and as adjuvant treatment.

In the following we will present in detail the conditions of examination and summarise our results.

DOUBLE-BLIND *PHASE II* CLINICAL TRIAL

Clinical trials started in 1995, after the acceptance of the pre-clinical file, with the permission of the National Institute of Pharmacy (Országos Gyógyszerészeti Intézet). Within two years, 42 patients were involved in the evaluation, from which the results of 35 were suitable for evaluation. After breaking the codes we stated that 19 patients belong to the treated and 16 to the control group.

The results of the evaluation are as follows. From the beginning of 1998, clinical trials were continued and a further 22 patients were involved till May 1999.

THE EFFECT OF DEUTERIUM DEPLETION ON TUMOURS OF THE PROSTATE INTERIM REPORT ON *PHASE II* CLINICAL TRIALS

GÁBOR SOMLYAI Ph.D., ANDRÁS KOVÁCS M.D., TIBOR KÁZMÉR M.D., IMRE GULLER M.D., JORGOSZ SZAPANIDISZ M.D., GÁBOR CSÚSZ M.D., GÁBOR ÁRPÁSI M.D., BÉLA BÖLCSKEI M.D., PÉTER JUVANCZ M.D., ISTVÁN JÁNOSI, GYÖRGY JÁKLI Ph.D., GÁBOR JANCSÓ Ph.D., GÉZA WABROSCH M.D.

SUMMARY

The aim of the trial: To prove that deuterium depleted water can effectively be used for the treatment of prostate tumour patients.

Methods: Double blind controlled *Phase II* clinical trial, in compliance with GCP principles.

Results: Interim evaluation confirmed a significant difference between the control and treated groups with respect to the examined parameters that indicated the anti-tumour effect of the preparation.

  a) At the time of the 5th and 6th visits, the ratio of patients showing an increased efficacy (PR) was significantly higher statistically (5th visit: $p = 0.0096$; 6th visit: $p = 0.021$) in the treated group.
  b) The volume of the prostate decreased significantly ($p = 0.043$) in the treated group, whereas it could be regarded as unchanged in the control group.
  c) The number of patients with a decreased prostate volume was significantly higher (exact Armitage-test: $p = 0.015$; exact Fisher-test: $p = 0.011$).
  d) Significantly more patients reported a positive change in symptoms in the treated group (exact Armitage-test: $p = 0.0009$; exact Fisher-test: $p = 0.0018$).
  e) The survival rate in the treated group was significantly higher ($p = 0.030$).

After the consumption of more than 10 tons of Dd-water no event endangering life happened. We did not experience any deterioration in blood counts, irritation of the mucous membrane, nausea, headache, etc., that could have been attributed to Dd-water consumption.

Conclusions: The application of water with a decreased deuterium content and the subsequently occurring deuterium depletion in the patient's body may be a suitable means of treatment of prostate tumours.

INTRODUCTION

During the past years, several independent experiments have proved the supposition that naturally occurring deuterium (HDO-concentration 16.8 mmol/L, 150 ppm) plays an important role in regulating intracellular processes. The first *in vitro* experiments revealed that if cells are placed in a medium of decreased deuterium content (90 ppm), their proliferation is inhibited for 10–12 hours [1]. The application of deuterium depleted water (Dd-water) resulted in tumour regression in mice transplanted with human breast and prostate tumours, as well as in dogs and cats with spontaneous tumours [1–6]. According to trials, Dd-water can trigger apoptosis *in vitro* and *in vivo* [6, 7].

The effect of deuterium depletion on biological systems was proved with regard to plants as well. Beyond its anti-tumour effect, Dd-water inhibited the growth of plant tumours [8] and plant sprouts of germinating seeds [9, 10] and it also activated membrane transport processes [4].

After the conclusion of pre-clinical and toxicological evaluations, human *Phase II* double blind clinical trials began in the summer of 1995. The aim of clinical trials was to prove the anti-tumour effect of Dd-water on tumours of the prostate. Here we will analyse the interim evaluation of the *Phase II* clinical trial.

According to the interim evaluation, prostate tumours can be treated successfully with Dd-water; the corroboration of this statement is presently going on with the involvement of further patients.

MATERIALS AND METHODS

Clinical trials are being carried out according to GCP principles with the permission of central and local ethical committees as well as the written consent of patients, simultaneously with conventional treatment. All patients involved in trials are receiving standard conventional treatment as well.

According to the protocol, only patients with a histologically confirmed prostate tumour can be involved. The duration of the trial is four months but the follow-up of the patients is continued even after the closing of the data files. During the trials the same two doctors followed the changes in prostate volume by a monthly rectal sonography. We calculated the volume of the prostate knowing the two diameters on the basis of the following formula: $V = 0.523 \times [(a+b)/2]^3$. During comparative analysis, the values measured at the 6th visit were expressed in the percentage of the value measured during the 1st visit, and the percent values thus obtained were compared in the two groups. Beside the change in prostate volume, several other parameters were recorded during the 4 months of follow-up (ECOG, urination problems, PSA, etc.).

Patients treated with the preparation consumed Dd-water of a 90 ppm D-content, while that of the placebo was 150 ppm. The daily dose, depending on the state of the patients, varied between 0.014–0.030 kg Dd-water/body mass kg, which corresponds to the consumption of a daily 1.2–1.8 L in the case of a patient with an average body mass (60–70 kg).

Evaluation of the results was done by the exact Armitage- and Fisher-tests.

Statistical analysis as to efficacy was based upon the comparison between the PR, NC, and PD case numbers in the two groups.

RESULTS

In the course of two years we treated 42 patients with prostate tumours. Three of them backed out before the beginning of the treatment, thus 39 patients (Stafety Population, SP) began to consume the coded preparation. Prior to breaking the codes, because of incorrect inclusion discovered later, a further 4 patients were excluded, thus interim evaluation was carried out with the inclusion of 35 patients (Intention-To-Treat, ITT), from among which 26 belonged to Group PP (Per Protocol: the patients could participate in all control examinations). With knowledge of the codes we learnt that 19 patients belonged to the treated and 16 to the control group.

Data of homogeneity of the treated and the control groups: Prior to the evaluation of results, we established that with respect to the phases of the illness, the two groups were homogenous. Evaluating the whole group together, there was no difference in the length of time between the setting up of the diagnosis and the involvement in the clinical trial (the mean value in the treated group was 294 days and in the control group 281 days). There was, however, a difference in the number of patients diagnosed within 3 months. The significance of this will be dealt with later. According to histological classification, 5 patients were treated with anaplastic adenocarcinoma, 4 of them belonged to the controls and only one to the treated group.

Change in the size of the prostate: According to the statistical analysis, the differences in the relative change of the size of the prostate (p = 0.043) deviated significantly in the treated and control groups (the mean value in the treated group was 78.8%±23%; and in the control group it was 97.9%±28.2%, in percent of the initial value).

Numerically this means that in the treated group a net 150 cm^3 decrease of the prostate volume occurred during 4 months of the treatment, while in the control group there was no net decrease in volume. The latter may be explained by the fact that although there were patients with whom the volume of the prostate decreased, this was compensated for by a significant increase with other patients.

The effectiveness of the preparation is also indicated by the fact that in the treated group significantly more patients experienced a decrease in the volume of the prostate (exact Fisher test: p = 0.028).

Table II. 9 *Changes in the volume of the prostate in the treated group*

Patient no.	V_1 (cm³)	V_2 (cm³)	V_3 (cm³)	V_4 (cm³)	V_5 (cm³)	Dir. of change (-, 0, +)	ΔV (cm³)	$\Delta\%$
1	55	45	49	32	37	–	–18	–32
4	74.5	43.6	–	32.7	26.8	–	–47.7	–64
6	29.1	–	27.9	29.1	23.7	–	–5.4	–18
7	20.8	–	13.6	–	15.8	–	–5	–24
9	19.9	15.0	12.9	1.8	6.0	–	–13.9	–70
11	–	–	–	–	–	–	–	–
14	78.9	66.3	–	–	74.5	–	–4.4	–5.6
17	20.8	20.8	19.0	19.9	18.2	–	–2.6	–12.5
19	31.4	29.1	31.4	30.2	29.1	–	–2.3	-7
21	35.2	35.2	32.6	33.9	32.7	–	–2.5	–7.1
24	56.8	49.9	–	–	–	–	–6.9	–12
26	46.7	26.8	27.9	–	24.7	–	–22	–47
27	32.6	32.6	30.2	32.6	30.2	–	–2.4	–7.4
29	60.5	–	62.3	60.5	66.2	+	+5.7	+9.4
31	11.6	11.6	6.0	9.3	–	–	–2.3	–20
34	14.3	–	15.8	–	14.3	0	0	0
36	30.2	35.2	32.7	19.0	14.3	–	–15.9	–52.6
39	19.1	–	17.4	–	–	–	–1.7	–9
41	33.9	33.9	33.9	–	–	0	0	0

The significant difference may be attributed to the fact that in the treated group, from among 19 patients only in one could the increase of 5.7 cm³ of the prostate volume be measured, which remained under 10 percent. In two cases

100

no change in the prostate volume was observed, while in the rest (16 patients), the volume of the prostate decreased. With 10 patients, the range of decrease was above 10 percent (12–70 percent). In contrast, in the control group, in 5 patients an increase in volume was observed (in 4 cases this was above 20 percent), in 2 cases it remained unchanged, and it was only in 9 cases that the volume of the prostate decreased (in 5 cases above 10 percent). The figures are given in *Tables II. 9–10.*

Table II. 10 *Changes in the volume of the prostate in the control group*

Patient no.	V_1 (cm³)	V_2 (cm³)	V_3 (cm³)	V_4 (cm³)	V_5 (cm³)	Dir. of change (-, 0, +)	ΔV (cm³)	Δ%
2	155.9	98.1	–	–	–	–	−57.8	−37
3	22.7	17.4	29	23.7	22.7	0	0	0
12	58.6	43.6	36.5	36.5	–	–	−22.1	−38
13	114.5	114.5	–	–	–	0	0	0
15	29.1	29.1	23.7	22.7	24.7	–	−4.4	−15
18	24.7	–	18.2	25.8	22.7	–	−2	8
20	15.8	16.5	18.2	19	–	+	+3.3	+20.1
22	30.2	–	20.8	–	18.2	–	−12	−40
23	39.3	18.2	18.2	40.7	40.6	+	+1.3	+3.3
28	26.8	25.7	23.7	27.9	25.8	0	−1	−3.7
30	14.3	–	17.4	–	–	+	+3.1	+21.7
33	11.6	–	–	4.6	11.0	0	−0.6	−5
35	51.5	–	78.9	–	74.5	+	+23	+44.6
38	62.4	60.5	85.8	–	–	+	+23.4	+37.5
40	–	–	–	–	–	–	–	–
42	22.7	–	9.9	–	9.3	–	−13.4	−59

Changes in urination problems: During statistical evaluation we analysed the data according to treatment groups, rendering observations into two groups: "OK" and "not OK". In the course of this we found that the changes in urination problems differed in the two treatment groups: the opinion of the change in the symptom was significantly better in the treated group (exact Fisher-test: p = 0.0018). Broken down to patients, numerically this meant that in the treated group from 19 patients 15 had urination problems at the beginning, and 9 of them reported an OK status, with no deterioration whatever.

In the control group, from the 10 patients who had urination problems at the beginning, only one reported having no problems, whereas 3 experienced deterioration (p = 0.69).

Evaluation of the efficacy of treatment: During the 4 months of the trial no CR was recorded, although with one patient who consumed Dd-water, no area poor in echo was detectable during sonography. The lack of CR cases, in our view, is due to the short period (4 months) of the treatment.

The breakdown of cumulative efficacy indexes on the occasion of the 5th and 6th visits in the ITT group is demonstrated in *Table II. 11.*

Statistical evaluation shows that the ratio of patients showing improvement at the 5th and 6th visits was significantly higher in contrast to the control group.

Table II. 11 *Breakdown of cumulative efficacy indexes at the 5th and 6th visits in the ITT group*

	Treated (5th visit)	Control (5th visit)	Treated (6th visit)	Control (6th visit)
PR	7	0	8	1
NC	9	9	6	8
PD	2	4	4	5

Breakdown of average survivals and deaths according to treatment groups: The 4 months follow-up prescribed in the protocol did not prove to be sufficient for obtaining appreciable results regarding survival, we therefore followed patients involved in the trial launched in August 1995 till May 1997. The average survival was determined as the length of the interval between the entrance into the trial as starting point, and the follow-up examination subsequent to the closing of the clinical trial as closing point.

According to statistical evaluation, average survival was significantly longer (p = 0.030) in the treated (ITT) group.

During the 4 months of participation in the protocol none of the patients died. After the closing of the trial, 7 of the 35 patients involved died, 2 of them from the treated group and 5 from the control group. Including the SP group as well, the results are demonstrated in *Table II. 12.*

Evaluation of laboratory tests and tolerance: In the case of laboratory tests there was a significant difference regarding gamma-GT and ALP: decrease occurred in the treated group, and increase in the control group (gamma-GT: p = 0.028; ALP: p = 0.022).

From the point of view of undesirable events we found no difference. According to doctors, the patients who consumed Dd-water, tolerated the preparation administered during the clinical trial significantly better (p = 0.024).

Table II. 12 *Breakdown of mortality in the treated and control groups*

	Treated	Control
SP group (19:20 patients)	2 (10.52%)	7 (35.00%)
ITT group (19:16 patients)	2 (10.52%)	5 (31.52%)
PP group (15:11 patients)	1 (6.66 %)	2 (18.18%)

DISCUSSION

During the past years numerous pre-clinical results have proved that when applying Dd-water, the subsequent decrease in deuterium concentration has an anti-tumorous effect in living organisms. The interim evaluation of *Phase II* clinical trials with tumours of the prostate supports our results so far, for in the most essential parameters (prostate volume, urination problems, efficacy, survival), results were significantly better in the treated group than in the control group.

It is a well-known fact that with conventional treatment the best results can be achieved in the period after the beginning of the treatment. We thus investigated separately, how many of the 8 patients in the treated group, and diagnosed within 3 months, were qualified as PR. Results showed that there was no link between the diagnosis of the disease and the cases qualified as PR, because from the PR cases 2 patients had been diagnosed almost a year earlier, one a year and a half earlier and another two and a half years earlier, and only the further 4 patients belonged to those freshly diagnosed. Thus, the large number of PR cases in the treated group cannot be attributed to the great number of freshly diagnosed cases, as the rate of these in the PR group was only 50 percent, similarly to that in the treated group of 19, the rate of freshly diagnosed was also nearly 50 percent (19/8).

There was heterogeneity in the breakdown of patients diagnosed with anaplastic adenocarcinoma. In the control group, a worse prognosis of anaplastic adenocarcinoma manifested itself, as from the 4 patients 3 were qualified as PD during the trial, and only one got a NC qualification. As opposed to the above, among patients in the treated group an unequivocal improvement was found in several parameters, but because of the low number of patients, evaluation did not take this into consideration.

Clinical trials are being carried out at present, too. Our aim is to reinforce the results achieved so far by increasing the patient number, and to prove, according to the conditions of GCP, that deuterium depletion causes tumour regression in the patients. Our future aim is to extend clinical trials to other types of tumours as well, and to prove the anti-tumour effect of Dd-water in the course of a longer follow-up.

REFERENCES

1. Somlyai, G., Jancsó, G., Jákli, Gy., Vass, K., Barna, B., Lakics, V., and Gaál, T. (1993): Naturally occurring deuterium is essential for the normal growth rate of cells. *Febs. Lett.* **317,** 1-4.
2. Somlyai, G., Jancsó, G., Jákli, Gy., Laskay, G., Galbács, Z., Kiss, A. S., and Berkényi, T. (1996): A csökkentett deutérium-tartalmú víz biológiai hatása [The biological effect of deuterium depleted water]. *Természetgyógyászat* **10,** 29-32.
3. Berkényi, T., Somlyai, G., Jákli, Gy., Jancsó, G. (1996): Csökkentett deutérium-tartalmú (Dd) víz alkalmazása az állatgyógyászatban [The application of Dd-water in veterinary practice] *Kisállatorvoslás* **3,** 114-115.
4. Somlyai, G., Laskay, G., Berkényi, T., Galbács, Z., Galbács, G., Kiss, S. A., Jákli, Gy., Jancsó, G. (1997): Biologische Auswirkungen von Wasser mit vermindertem Deuteriumgehalt. *Erfarhrungskeilkunde* **7,** 381-388.
5. Somlyai, G., Laskay, G., Berkényi, T., Galbács, Z., Galbács, G., Kiss, S. A., Jákli, Gy., Jancsó, G. (1998): The biological effects of deuterium-depleted water, a possible new tool in cancer therapy. *Z. Onkol./J. of Oncol.* **30,** 91-94.
6. Somlyai, G. (1998): Csökkentett deutérium-tartalmú víz – Új lehetőség a daganatterápiában [Deuterium-depleted water – A new possibility in cancer therapy]. *Komplementer Medicina* **2,** 6-9.
7. Somlyai, G., Laskay, G., Berkényi, T., Jákli, Gy., Jancsó, G., (1997): Naturally occurring deuterium may have a central role in cell signaling. In: Heys, J. R. and Melillo, D. G. (eds.) *Synthesis and Applications of Isotopically Labeled Compounds.* John Wiley and Sons Ltd. 137-141.
8. Kiss, A. S., László, I., Szőke, É., Galbács, Z., and Galbács, G. (1997): The effect of deuterium depleted medium on plant tumours. In: Theophanides, T. and Anastassopoulou, J. (eds.) *Magnesium: current status and new developments.* Kluwer Academic Publishers, 81–84.
9. Kiss, A. S., Galbács, Z., and Galbács, G. (1997): Magnesium ions hinder the growth of coleoptyl in deuterium depleted water. In: Theophanides, T. and Anastassopoulou J. (eds.) *Magnesium: current status and new developments.* Kluwer Academic Publishers, 77–80.
10. Belea, A., Kiss, A. S., Galbács, Z. (1997): Eltérő módon asszimiláló növények C3, C4 és CAM típusának meghatározási módszerei [Methods of defining types C3, C4 and CAM of plants]. *Növénytermesztés* **46,** p. 477-485.

Abbreviations: SP – Safety Population; Dd-water – deuterium depleted water; ITT – Intention-to-Treat Population; PP – Per-Protocol Population; GCP – Good Clinical Practice; CR – Complete Remission; PR – Partial Remission: NC – No Change; PD – Progressive Disease; ALP – Alkaline Phosphatase.

APPLICATION OF DEUTERIUM DEPLETION ON THE BASIS OF COMPASSIONATE USE AND AS ADJUVANT TREATMENT

Prior and parallel to the above trials, from October 1992 till the spring of 1999, we provided Dd-water for approximately 1200 patients. Our knowledge concerning the efficacy and application methods of Dd-water comes mainly from the follow-up of this patient population. Their provision with Dd-water and their follow-up occurred under the following conditions:
- patients were given Dd-water beside conventional treatment and not instead of it;
- we stressed the importance of the patients' informing their doctors about the consumption of Dd-water, and aimed at co-operation with the attending physician;
- patients involved were given a code and their data recorded in a protocol;
- this enabled us to draw up a detailed and authentic statistical analysis of the patient population consuming Dd-water;
- during the follow-up we recorded the data of medical control examinations, the quantity and concentration of Dd-water consumed;
- we recorded all the experiences and observations of the patients thought to be related to Dd-water consumption.

With regard to the chemical composition of the preparation (water with a deuterium content lower than natural), and to the proved relative harmlessness of it, as well as the results of the interim evaluation of the *Phase II* clinical trial, from January 1998, with the approval of the National Institute of Pharmacy, hospitals also joined in the application on the "trial principle" of Dd-water. (The "compassionate use" means that the future drug under approval can be given to patients with whom, during treatment, medical science had exhausted all resources at its disposal, so they "attempt" treatment with a preparation whose clinical trials are still being carried out to fully prove efficacy.)

This group of patients is significant due to three factors:
a) The patient population is relatively large, the number of patients involved is around one thousand, the number of appreciable cases, who have been consuming Dd-water for a minimum of 3 months, is 500–550.

b) As the first patient was involved 8 years ago, we have a 2–5 year overview on the patient group of several hundreds.

c) The patient group represents a variety of tumorous diseases.

When evaluating the cases the absence of the control group may, of course, be objectionable. In our opinion, the control group is made up of the 210 thousand cancer patients who have died in Hungary in the past seven years, as well as the further 200–250 thousand who became ill in the last 4–5 years and were not given Dd-water treatment. The efficacy of Dd-water is proved by the fact that we can with great frequency show case descriptions with a significantly different course from the average. In the control group where patients were given conventional treatment only, from among 10 thousand patients it is only one with whom an unexplainable improvement or spontaneous healing is to be expected (0.01 percent). Among those who consumed Dd-water, 10 percent showed an outstandingly favourable disease process. This means that among the Dd-water consumers the number of the unexpectedly positive cases is 1000 times greater than in the average patient population.

It is most regrettable that these results are often dismissed as simply "anecdotes". The very logic of this approach is hard to understand. With the researcher's eye it is exactly the exceptional that should be welcomed. One should try to find out the cause, as this leads to the solution. There must be an explanation for somebody's improvement or recovery, or for the fact that his state does not worsen as expected. Not to take these cases into consideration shows an irresponsible and superficial attitude.

Certain tumour types are represented in a nearly similar ratio among the more than 1000 patients and in the entire patient population. From this it follows that in the case of frequently occurring tumour types (colon, lung, breast) we may speak of a patient number above one hundred. In another group, 20–60 patients were treated per tumour type, while in the case of rare tumour types, the data of only some patients are at our disposal.

During the last 8 years we provided about 350 tons of Dd-water for the patients, and some 12–14 thousand pages of documentation records the data of the meticulous follow-up. Recommendations, comments, dosage advice, and results are based on these observations.

The present chapter introduces our results based on detailed statistical evaluation, and observations made with breast cancer patients. We would like to stress that treatment was performed mainly with 90-ppm Dd-water, which was at our disposal in great quantities. Our results described here were significantly surpassed when, from the summer of 1998, Dd-water of a D-content lower than 30 ppm was at our disposal.

Based on case studies of certain patients treated successfully, we wish to present the efficacy of Dd-water with further tumour types as well. Thus an overall picture can be presented of the possibilities of Dd-water application. We are convinced that many case descriptions may even seem unbelievable. We do not expect anyone to believe our results; we consider it important that doctors themselves could be convinced of the correctness of our statements. Results presented here are easy to reproduce with further patients as well.

We are fully aware of the fact that the following data need further amendments and corrections. We do not aim to hand over the experiences of a *Phase III* clinical trial with the involvement of several hundred patients, as our drug development is not in that phase yet. By presenting the case descriptions, we would like to call attention to the simple fact that in the majority of cases, a significant positive change was observed in the state of the patients following the start of Dd-water consumption, even if their condition had progressed further beside conventional treatment. This simple observation encouraged us with good reason to consider their improvement as the consequence of deuterium depletion, and to see a cause and effect relationship between the consumption of Dd-water and the subsequent improvement of the patient's health.

DETAILED ANALYSIS OF THE DATA OF BREAST TUMOUR PATIENTS

In the following we will perform a retrospective analysis of the patient population with breast cancer (the period between October 1992 and December 1997). Evaluations refer only to patients who had, within this period, consumed Dd-water for a time longer than three months. Our results reflect the situation before July 1998.

With the exception of several cases, patients were given conventional treatment, too. Results, therefore, should be considered as the joint effect of conventional and Dd-water treatment.

Between October 1992 and December 1997, 887 patients had begun to consume Dd-water. Among them 134 patients (15 percent) were diagnosed with breast cancer. This ratio shows that patients with breast cancer were represented in an approximately equal ratio among patients consuming Dd-water, to that of the entire patient population (in the US, for example, 13 percent of all cancers is that of the breast).

The patients were initially divided into two main groups. The first consisted of those patients whose results we wished to use, while the second group were patients who had been consuming Dd-water for a period shorter than three months. The division of patients in the two groups was 51–49 percent, which means that the data of 68 patients were evaluated and 66 patients were excluded from evaluation.

We first wish to give a short survey of the patients excluded from evaluation.

Breakdown of patients consuming Dd-water and excluded from evaluation:

Weeks	0–4	30 patients
	5–8	11 patients
	9–12	25 patients

Seventeen patients died within the 3 months period (25.8 percent), 3 after that (4.5 percent). Thirty-five patients were living after finishing the Dd-water cure (53 percent), and no information was available about 11 patients (16.7 percent).

The data of the patients involved in evaluation are given in *Table II. 13*. We indicated the serial number in the protocol of the patient, the time between the diagnosis and the beginning of the Dd-water protocol (Diagn.–DDW protocol), the length of the Dd-water protocol, and the extent of survival. (In some cases the extent of survival exceeds the sum of the time span between the diagnosis and the beginning of the cure plus the length of the Dd-water protocol. This is explained by the fact that although the patient was no longer consuming Dd-water, we followed her even afterwards.) We recorded who of the patients had died and in what condition they had begun to consume Dd-water (R: patient in remission; P: the primary tumour had been diagnosed directly before the beginning of Dd-water consumption; A: recidiva, the appearance of distant metastases.

Table II. 13 *Main data of breast cancer patients consuming Dd-water*

Diagn.-DDW protocol (weeks)	Dd-water protocol (weeks)	Survival (weeks)	Died (+)	Code	Diagn.-DDW protocol (weeks)	Dd-water protocol (weeks)	Survival (weeks)	Died (+)	Code
52	42	94	+	A	104	34	138		A
4	25	37		A	104	52	156		A
245	252	497	+	A	0	130	130		P
49	29	200	+	A	572	34	606	+	A
91	104	364		A	0	48	48		R
20	29	186		A	0	52	200		P
77	54	131		A	234	52	286	+	A
81	56	206		A	38	56	168		R
56	33	116	+	A	234	34	268	+	A
52	16	68	+	A	226	132	358	+	A
4	64	89		A	222	52	274		A
12	32	44		A	1040	148	1188	+	A
22	20	104	+	R	4	164	168		A
234	56	312		A	108	164	272		A
52	66	118		A	16	29	77		A
36	90	144		R	8	56	64		A
468	108	576	+	A	296	26	322		A
364	13	377		A	0	160	160		P
572	25	597		A	476	152	638		A
21	142	182		A	12	144	156		A
96	21	117	+	A	12	144	156		R
676	52	728		A	12	34	116		P
56	100	156		R	138	125	263		A
30	47	77		A	0	134	134		P
364	53	417	+	A	312	51	363	+	A
156	69	225		A	34	65	99	+	A
21	69	90		A	4	100	104		A
12	168	180		A	0	100	100		P
408	200	616		A	177	86	263		A
3	100	103		R	546	26	572	+	A
69	212	281		A	454	64	518	+	A
82	60	256	+	A	728	69	797		A
364	52	416		A	356	17	373		A
0	45	204		A	338	38	376		A

Note: Sixty-eight patients had been consuming Dd-water for altogether 5 276 weeks.

Statistical evaluation covered 55 patients in an advanced state (A).

Figure II. 4 shows the survival data of breast cancer patients in an advanced state and consuming Dd-water.

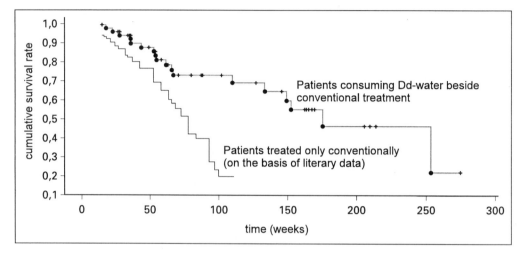

Figure II. 4 *Survival graph of breast cancer patients*
treated after the appearance of distant metastases

It is clear from *Figure II. 4* that 85 percent of the above population survived one year after the beginning of Dd-water consumption in an advanced state, 56 percent survived 3 years, 47 percent 4 years, and 25 percent 5 years after the beginning of consumption of Dd-water.

Experiences and statistical data gained during the past decades may serve as controls. According to these, in the case of metastases in the liver, lung, bone, and central nervous system, life expectancy in this patient population is around 12–18 months (52–78 weeks). In cytostatic treatment, which is considered to be the most efficient, the fact that 20 percent of the patients survive for 2 years and the average 18 months survival rate grew, is considered to be a significant breakthrough.

In the patient population consuming Dd-water, 73 percent of the patients survived for 2 years after the appearance of distant metastases and there was a 25 percent chance of a 5-year survival. These results confirm our earlier data according to which tumour patients can, with great efficacy, be treated and cured.

To represent what changes the application of Dd-water induces in patients with breast cancer; we have prepared short case descriptions of some patients. These examples serve as direct proofs for the anti-tumour effect of Dd-water; connections are evident, and improvement and survival rates by far exceed statistical values that could be attributed to "spontaneous" improvement.

Female patient (46); on Dd-water from 12-07-1993
(Uzsoki Utcai Kórház, Budapest)

After the diagnosis of breast cancer in 1988, bone metastasis was verified in September 1992 for the first time. Until the patient started drinking Dd-water, examinations verified progression, her pains and disability increased. **Following the administration of Dd-water, her pains subsided within 1–1.5 months, bone scintigraphy made two months later could not verify several smaller metastases that had earlier been present.** The patient consumed Dd-water until January 1994 and noted an improvement in the quality of life. She then suspended the consumption of Dd-water. After four months she experienced significant deterioration, her pains reappeared. Bone scintigraphy, performed in September 1994, verified moderate progression. After this the patient decided to resume the cure. Bone scintigraphy of October 1995 could not reveal further metastases. Despite the above, there was a slow progression, while in the first three years of consuming Dd-water her quality of life did not change significantly. In December 1996, due to metastases, she suffered a pathological fracture in her upper arm, and her thighbones were surgically reinforced. The upper arm bone knitted, in the summer of 1997 the patient regained 10 kg of the lost bodyweight and her pains were controllable with drugs. During the first four years of treatment, no metastasis occurred in the soft tissues. CT scanning in October 1997 revealed brain metastasis. The patient died five and a half years after the appearance of the bone metastasis; she had been consuming Dd-water for four and a half years.

Female patient (39); on Dd-water from 23-08-1993
(Jósa András Kórház, Nyíregyháza)

Following surgery in October 1991, the patient received CMF–radiotherapy–CMF treatment. In July 1993, block-dissection was performed in her right armpit. She had been consuming Dd-water between August and November 1993. In the fall of 1994 she complained about double sight and CT scan verified an affected area in the temporal region. From December 1994, she had been consuming Dd-water irregularly, not in the recommended quantities. In the spring of 1995, mediastinary and pulmonary affection was verified. The patient resumed the consumption of Dd-water between March and September 1995, later stopped and started again in October 1996. She has been drinking Dd-water ever since continuously. Her state reflected stagnation, progression and improvement alike. **The patient is still alive (May 1999), despite her diagnosis of brain metastasis and lung metastasis five years earlier.**

Female patient (72); on Dd-water from 11-03-1994
(Uzsoki Utcai Kórház, Budapest)

The patient underwent breast surgery in 1985. At the end of 1993, examinations verified metastases in the brain, liver, and bones. She had been consuming Dd-water from the above date till her death in May 1996. In contrast to the expected 2–3 month of survival, she lived for 26 month in a stable and fairly good general condition.

Female patient (39); on Dd-water from 03-09-1994
(Területi Kórház, Berettyóújfalu)

The patient underwent surgery first in 1986, when a walnut size tumour, that had been present for a year then, was removed. The first recidiva appeared in 1987. The suspicion of bone metastasis came up in September 1993, verified by examinations in March 1994. The illness progressed despite intense conventional treatment. Shortly before the start of Dd-water consumption the following is to be read in her final report: "St. P. amput. Mammae l. d. pp. Cc. Metastasis localis cutis. Metastasis ossis capitis. Metastasis vertebrae C. III. proc. Spin. Th. Iv. VII. VIII. IX. XII: L. I. II. III. Metastasis pulm. L. s. Metastasis costae V. l. d. ilei l. d. Laesio plexus brachialis l. d. Anaemia sec. St. P. ovariectomiam..." **Because of her severe pain, the patient needed morphine. Some weeks after the start of the cure she no longer needed painkillers.** In October, a tumour under her nose became flatter and smaller in size. **X-ray examination in November verified the regression of the lung metastasis. Bone scintigraphy in March 1995 also showed regression. In January 1996 examination stated full regression in the lung**, with further regression in the bones. Metastases under the skin of the head further progressed from November 1996. By January 1997, the lungs were clear, the patient had no pain. She drank Dd-water till the spring of 1998 and after a half-year break she resumed the consumption of Dd-water in November 1998. Subsequent bone scintigraphy described minimal metastases at some points only. The patient has been free from pain for five years, since the beginning of the Dd-water cure, the quality of life is good, and she leads an active life.

Female patient (56); on Dd-water from 10-01-1995
(SOTE Radiológiai Klinika, Budapest)

In August 1990 the patient underwent mamma exstirpatio, histology proved invasive cc. mammae. After surgery the patient was given radiotherapy. In 1992, metastasis appeared in her left breast, too. Prior to the consumption of Dd-water, examinations verified metastases in the ribs and in the L. II–III–IV vertebrae. The patient had severe pain in the back. Three months after the start of the Dd-water cure, her complaints eased, and her general condition improved. Blood counts of June 1995 showed considerably better results, ESR value instead of the former 60 decreased to 26, no purulence appeared in the urine. By October the condition of the patient had further improved, she gained weight. Influenza in February 1996 retarded improvement for a long time. In August, examinations confirmed a metastasis in the brain. As a consequence of both Dd-water and conventional treatment, intermittent improving and worsening conditions followed until June 1997, the death of the patient. There was an unanimous improvement after the consumption of Dd-water, and a 30 months survival after the appearance of bone metastasis and 10 months survival after the diagnosis of brain metastasis.

Female patient (42); on Dd-water from 18-05-1995
(Országos Onkológiai Intézet)

Following breast surgery in April 1993, the patient received radiotherapy. In March 1995, local recidiva appeared as well as bone metastases in the vertebrae and the sternum. The patient began consuming Dd-water two months later, and had been consuming it for 42 months. Bone scintigraphy made in December 1995 showed no progression. The pains subsided, and the general state of the patient improved. The year 1996 elapsed in a good general health. In January 1997 pulmonary oedema appeared, but apart from occasional pain in her sacrum, the patient felt well. X-ray scan performed in October 1997 reinforced suspicion of a lung damage but the former ovarian cyst could not be verified. In January 1998 the patient began to cough, and examinations verified lung metastasis, after which the cytostatic treatment of the patient began.

Female patient (48); on Dd-water from 28-08-1995
(Országos Onkológiai Intézet)

Because of **invasive lobular cc**, in the summer of 1986, sectoral excision was performed from the left breast with axillar block dissection. After surgery the patient was given radiotherapy but she would not accept adjuvant chemotherapy. In 1990 recidiva were removed from the left breast, followed by radiotherapy. In August 1995, wide inoperable recidiva appeared infiltrating the whole breast. Then a combined cytostatic treatment began after the CMF-scheme. The consumption of Dd-water also started in August 1995. **The subsequent regression was more than 50 percent. After finishing chemotherapy regression continued.** The patient has been drinking Dd-water since August 1995, is free of complaints and returned to work.

Female patient (37); on Dd-water from 26-02-1996
(Országos Onkológiai Intézet, Budapest)

The patient had first been operated with breast tumour in 1993, after which she was given radiotherapy. The bone scintigraphy performed in 1995 verified metastasis; the chemotherapy begun afterwards lasted till October 1996. Because of the cytostatic treatment the patient consumed Dd-water irregularly between February and August 1996, but after several breaks in the cure, she has been consuming it regularly since August 1996. **In the period between August and October 1996, the patient gained 3 kg of weight, and needed no help with walking.** She has been active ever since and has no pain. According to the 1998 control, the process has consolidated; a significant calcification was verifiable in the bones. The constriction in the hips has loosened; **the patient tolerates significant physical stress (swimming, mountain climbing)** and is in good general condition.

Female patient (48); on Dd-water from 19-11-1996
(Péterfy Sándor Utcai Kórház, Budapest)

Anamnesis indicated a right side breast ablation, because of a tumour in the T2N0M0 state. The patient was given adjuvant radio-chemotherapy during surgery. In August 1996 pleural and lung metastases were verified. **Cytostatic treatment started according to the MMM scheme, from which 4 series were administered until the start of the Dd-water cure. Control examinations did not reveal any changes in the size of the tumour. The patient started to consume Dd-water at this point. By January 1997, the size of lung metastases diminished considerably, the patient gained weight. According to the CT scan of March 1997, a significant regression took place in the lung. In August 1997 complete regression was reported,** and MMM treatment was finished. In the fall of 1997, the patient was able to travel abroad and gave a course outside the country. The X-ray examination of October showed further improvement, a residual tumour remained in the pleura only. The control examination of February 1997 gave a negative result. In July 1998, a progression of the metastases in the pleura was observable. Taxotere treatment began in August 1998, which was finished in November.

Female patient (54); on Dd-water from 18-03-1997
(Uzsoki Utcai Kórház, Budapest)

The patient underwent her first operation in 1983, during which the involvement of the lymphatic nodes was verified. In 1992, after an epileptic seizure, metastasis in the brain was revealed. After surgery the patient was given chemotherapy and radiation. In August 1996, metastases in the lungs appeared. CT scan in March 1997 described, beside the metastasis in the lungs, in the liver and adrenal gland as well. **In August 1997, five months after the beginning of the Dd-water cure, the liver metastasis was stagnating, in October it showed regression, and in June 1998, complete tumour regression was verified. The abdominal sonography in October revealed a negative result with regard to both the liver and adrenal gland.** By the end of 1997 deterioration was experienced, and the patient was given radiotherapy followed by two doses of cytostatic treatment. Since April 1998, the patient has been in a good physical state with no complaints. During the summer she coughed up considerable amounts of white jelly-like secretion which is a characteristic phenomenon following Dd-water consumption of patients with the involvement of the lungs.

EXPERIENCES WITH OTHER TUMOUR TYPES

In the following we present experience gained with several other tumour types. By describing the cases we wish to demonstrate the efficacy of Dd-water on the one hand and highlight the characteristics of other tumours, on the other. We stress again that results presented here do not exhaust all possibilities of Dd-water treatment, because the bulk of experiences come from the time when, lacking Dd-water of a low D-content (30 ppm), it was not possible to maintain decreased D-concentration for a half year–one year period. (With 90-ppm Dd-water at our disposal, D-concentration in the patients' body decreased to 115–120 ppm.) We are convinced that the cases described below can be reproduced with medical assistance, and in the possession of newer clinical experiences, the efficacy of Dd-water can be further increased.

Tumours of the lung

> Male patient (62); on Dd-water from 21-04-1993
> (Budai MÁV Kórház, Budapest)

The condition was diagnosed in 1992; thoracotomy verified an inoperable tumour. After the diagnosis, the patient was given radiotherapy and first consumed Dd-water regularly between April 1993 and September 1993. During this time no progression could be traced. Later the patient gave up the cure because of the pain that appeared 5–10 minutes after drinking the water. Following this the patient lost 15 kg until 15 March 1994, when, on the advice of his doctor, he took up consuming Dd-water. In two months the patient gained 4 kg, according to the X-ray performed in July, the size of the tumour was corresponding to the state observed 2 years earlier. The patient's asphyxia stopped, he became physically active again. **The examination of January 1995 verified the stagnation of the tumour,** bronchoscopy executed in June described cicatrisation in the tumour area. After this, the patient was in good physical condition, **in May 1996 X-ray scan verified regression.** The patient stopped the consumption of Dd-water in a good general condition in August 1996. From January 1997 a renewed deterioration followed, in May 1999 the patient complained of pain in the operation area. Physically he was active and his weight stable. Seven years have elapsed since the diagnosis and the explorative operation, **according to the X-ray image, the tumour has encysted.**

Male patient (72); on Dd-water from 05-12-1993
(Tüdőkórház, Deszk)

The weakened patient had a relatively small (1 cm) tumour in the upper left lobe of the lung, near the artery wall, verified by both X-ray and endoscopy. After the first month of Dd-water consumption, the tumour showed a minimal growth but by January no growth could be verified. The patient up till then confined to bed was able to get up, his physical condition improved; he could walk for hours. The patient had been consuming Dd-water continuously till August 1995; later, in April and October 1996 and October 1997 he repeated the cure. According to medical reports of April 1996, the tumour had encysted. Because of his haemophilia, the patient was given no conventional treatment. According to information of October 1997, the patient's condition was stable, although 4 years earlier doctors defined his life expectancy to be several weeks only.

Male patient (47); on Dd-water from 10-01-1994
(Tüdőkórház, Deszk)

The patient had been treated from May 1993 on with inoperable lung cancer of an epithelial origin. Until the beginning of the Dd-water cure, the patient had lost 15 kg; the size of the tumour stagnated, later a minimal regression and, according to the December 1993 examination, progression was verified. Because of the progression, from January 1994, a change in the protocol was introduced. The patient tolerated the treatment well; his state was characterised by stagnation and an occasional decrease of atelectasia (August 1994). He got his last treatment in September 1994. **The control examination in December 1994, performed 3 months after the last cytostatic treatment when Dd-water was the only kind of treatment, described a "large-scale regression" that continued according to the examination results of March 1995.** Progression again was detectable in January 1996; the patient died in October 1996, three and a half years after the diagnosis of the illness. He had been consuming Dd-water continuously.

Female patient (54); on Dd-water from 15-02-1994
(Tüdőkórház, Deszk)

In January 1994, the patient had been diagnosed with **adenocarcinoma of the lung**. One month later she began to consume Dd-water and **had been drinking it continuously for 20 months,** until October 1995. **During this time no progression occurred**, in June 1995 the tumour that earlier had been verified as to be of the size of a man's fist, became smaller. The patient was in a good general condition when she stopped drinking Dd-water. According to our information of September 1998, **four and a half years after the diagnosis with lung adenocarcinoma, she was active and working**, the tumour had for years been stagnating after the consumption of Dd-water.

Female patient (46); on Dd-water from 29-09-1994
(SOTE Pulmonológiai Klinika, Budapest)

In March 1994, a tumour of 4 x 5 cm and one of 8 cm in size, of a mixed histological type (**planocellular + anaplastic adenocarcinoma**) was diagnosed. The beginning of the Dd-water cure coincided with a change in the protocol, thus the 50 percent decrease in the volume of the tumour experienced two weeks later can be considered to be the joint effect of the two kinds of treatment. Cytostatic treatment ended in February 1995. By March the patient had no more asphyxiation, her general condition was good, by June she gained 12-14 kg of body weight. A renewed progression was verified in August; the patient **died in March 1996, two years after the set-up of the diagnosis.**

Female patient (61); on Dd-water from 29-11-1994
(Tüdőkórház, Deszk)

The patient had been diagnosed with **microcellular lung cancer** in September 1994; cytostatic treatment began in October. The patient reacted well but tolerated the grave side-effects with difficulty. She had been consuming Dd-water continuously from November 1994 till October 1995; her physical state was satisfactory. She did not show up after this, but from a patient treated in the same hospital we know that she died **four years after the diagnosis and 3 years after finishing the Dd-water cure.**

Female patient (75); on Dd-water from 18-04-1995
(Országos Korányi TBC és Pulmonológiai Intézet, Budapest)

The patient was operated in August 1993 with adenocarcinoma. The tumour was attached to the wall pleura in the apex. In the centre of segment III, a tumour of the size of a walnut was found, this and the hilar lymphatic nodes were removed by lobotomy. The Ca 19-9 tumour marker value of the patient was increasing from the end of 1994:

November 1994:   54.8 ng/ml
January 1995:   54.3 ng/ml
February 1995:   74.9 ng/ml

In the first three months of consuming Dd-water the patient lost 5 kg, and the tumour marker further increased.

July 1995:   86.1 ng/ml

After the increase of the Dd-water dose, from December 1995 the tumour marker values were as follows:

December 1995:   60.8 ng/ml
March 1996:   45.4 ng/ml
October 1996:   39.4 ng/ml
January 1997:   29.6 ng/ml
June 1997:   23.5 ng/ml
January 1998:   20.2 ng/ml

The patient has been consuming Dd-water for the fifth year; lives in a good physical condition and without complaints six years after the operation.

Male patient (69); on Dd-water from 10-10-1995
(Szt. Ferenc Kórház, Miskolc)

In June 1994 a left side pulmonary neoplasia was verified (**planocellular**), which, during exploration, proved to be inoperable. **The tumour infiltrated the pericardium and entwined the main arteries.** Because of the size and location of the tumour, conventional treatment had to be rejected; thus, **Dd-water was the only means of treating the patient. After a six month Dd-water cure, by April 1996 the patient noted improvement in his health.** ESR fell from 80 to 6, the size of the tumour stagnated. The **blood sugar level** of the diabetic patient decreased, his dyspnoea lessened, the accumulated residues cleared up from the left lung, and the patient was able to return to his work as a physician. **In the fall of 1997, the patient felt well enough to undertake a journey abroad.** His state was stable until June 1998, when a slow tumour growth was observed. In July 1998 the patient underwent a myocardial infarction and had dyspnoea even at rest. Heart complaints were stabilised with medication, and his tumour showed a minimal growth only, despite the fact that following the heart attack he stopped the consumption of Dd-water, and restarted in January 1999 only. **In June 1999 the patient had no dyspnoea when resting and X-ray showed a moderate progression. He has been consuming Dd-water for 4 years now, with only one break in the meantime.** The patient regularly coughs up a considerable amount of secretion.

Note: With regard to the frequent occurrence of lung cancer, we have described several cases. It is clear from the above that we were able to provide positive examples to demonstrate all types of lung cancers. If we were to make a statistical summary, we would see that in 60–70 percent of the cases there was an objective reaction after the consumption of Dd-water; survival rates increased significantly and the tumour could, for years, remain dormant. The case of the 75-year-old patient illustrates well that in remission following surgery, Dd-water can prevent later recurrence, which would supposedly have happened taking tumour marker values into consideration. Results can significantly be improved by beginning the treatment with 75–85-ppm Dd-water instead of 90-ppm water. This is justified especially in the case of large-size (4–5 cm) tumours.

Colon, rectum

Female patient (55); on Dd-water from 03-11-1993
(Honvéd Kórház, Győr)

Sonography made in October 1993 verified **5 liver metastases** originating from the tumour of the rectum. **The patient complained of severe pain in the liver area, which stopped within days after Dd-water consumption.** In February 1994, the patient **had not been consuming Dd-water for a week, his pains reappeared but stopped again after the renewed consumption of Dd-water.** The patient's state was stable and he was active during the year 1994 and in early 1995. She stopped consuming Dd-water in March 1995, and **died in July 1995, 21 months after the verification of metastases in the liver.**

Male patient (65); on Dd-water from 10-10-1994
(BM Kórház, Budapest)

In September 1994, originating from a colon tumour, a 3–6 cm multiplex metastasis was verified in the liver. Tumour marker values were measured right before the beginning of Dd-water consumption (06-10-1994), then four months later (20-02-1995). The values were as follows:

CEA:	959 ng/ml	189.7 ng/ml
CA-50:	998 U/ml	195.8 U/ml
Ca-242:	3069 U/ml	324.6 U/ml

The patient had been consuming Dd-water until his death in July 1996. **The patient survived 22 months after the diagnosis in September 1994.**

Male patient (58); on Dd-water from 07-02-1995
(Országos Onkológiai Intézet, Budapest)

In February 1995, a 60-cm section of the colon was removed in surgery. **In the liver, 5–6 metastases were found, one of which was the size of an apple.** The patient was given 5FU + Leukovorin treatment. According to the control examination in March, after the consumption of Dd-water for a month, the two kidney cysts were not detectable, by May the patient gained 6 kg of weight. CT and sonography in October verified stagnation and the presence of calcified spots in the liver. CT scan made in August 1996 also verified stagnation. From October occasional diarrhoeas occurred, followed by significant deteriorating in health. The patient **died in May 1997, 27 months after the diagnosis of the illness and the appearance of metastases in the liver.**

Note: With this tumour type, in the case of distant metastases we "only" succeeded in significantly enhancing survival. This tumour type reacts to deuterium depletion very slowly, and 90-ppm Dd-water was not of a sufficiently low D-content to be able to eliminate larger metastases in the liver. It would be of great significance if, the patient consumed Dd-water continuously for 1–1.5 years after colon-rectum surgery, as a cure could prevent the appearance of distant metastases. With this tumour type and in the presence of distant metastases, it is advised to start Dd-water treatment with water of 75–62.5 ppm D-content.

Prostate

Male patient (65); on Dd-water from 30-10-1992
(Dél-pesti Kórház)

The patient was diagnosed with **inoperable prostate tumour** in October 1992. After the beginning of the cure he noted an improvement in urination problems, the PSA value decreased even prior to the start of the Fugerel treatment. **After one month the tumour became impalpable and also operable,** but the patient (a physician by profession) refused to undergo surgery. After drinking Dd-water for a year, the patient **has been free of complaints for seven years.**

Male patient (66); on Dd-water from 01-12-1994
(Szeged, MJVÖ Kórháza)

In September 1994 the tumour of the prostate caused total obstruction and metastases in the lymphatic glands. In October the **PSA value was 83.4 ng/ml.** Following Dd-water consumption, urination problems stopped within a short time. **Two weeks after the start of the Dd-water cure, blood samples revealed**

a 0.99 ng/ml PSA value. This, however, was thought to be erroneous, therefore **two weeks later it was repeated, and showed a value of 0.6 ng/ml.** This value decreased by March 1995 to 0.23 ng/ml, whereas values measured in June 1996 and January 1997 were below the traceable level. The patient had, for the first time, been drinking Dd-water continuously for 9 months, and repeated the cure for 2–3 months in 1996.

Note: In the case of prostate tumour it is of great significance that by measuring PSA value, the potentially tumorous patient population can be defined. Based on our results so far, we are of the opinion that if we are able to diagnose the illness in its early phase, with the application of Dd-water this tumour type can be cured with great efficacy, without the need for drug treatment throughout the patient's whole life. It would be of great economic significance to test the male population above 50–60 for PSA value and to treat cancer patients with Dd-water. For the time being, the cheap test (some thousand forints/test) is followed by a very expensive treatment (approx. 600–700 thousand forints/year/patient) which, in the majority of cases lasts throughout the entire life of the patient. Not even the United States of America can afford to provide preventive treatment for all men with tumour of the prostate, as this would cost about 100 billion dollars. The application of Dd-water would make the treatment followed by cheap scanning accessible for everyone.

Ovaries

Female patient (45); on Dd-water from 02-1992
(Szt. István Kórház, Budapest)

The patient was operated with tumour of the ovary in 1993. Following surgery, her tests were negative; she had good appetite, gained 4 kg and was active and working. She had been consuming Dd-water continuously between February 1993 and November 1994, when, with regard to the long symptom-free period, she stopped the cure. Results of examinations in December 1994 were negative, but two months later complaints in the abdomen appeared. In February 1995, a repeated operation took place. After this the patient resumed the consumption of Dd-water. It is to this that we attribute that despite the positive cytological result, sonography turned out to be negative again. In April 1996 recidiva could be verified. The patient died in November 1997, after a long struggle, nearly three years after the first relapse. (After this, we did not recommend to other patients operated with tumour of the ovary to stop drinking Dd-water even after a two-year-long symptom-free period.)

Female patient (52); on Dd-water from 04-04-1996
Péterfy Sándor Utcai Kórház, Budapest)

The patient was diagnosed with tumour of the ovary in the summer of 1995 which, histologically, proved to be a highly differentiated adenocarcinoma. **The uterus and the omentum were also affected.** After surgery the patient was given a series of 8 cytostatic treatments (Carboplatin, Cysplatin) that ended in March 1996. **She began to consume Dd-water after conventional treatment was over and has been drinking it for 40 months continuously – she is active and symptom-free.**

Note: It is a well-known fact that ovarian tumour is a type of a very poor prognosis. In spite of this, however, according to our experiences it can be well treated with D-depletion. We would like to stress that between prognoses based on our present knowledge and the efficacy of Dd-water there is no strict correlation. This means that tumour types that seem hopeless from the traditional point of view may react well to Dd-water treatment. In the case of tumours of the ovary we wish to highlight that in some cases a relapse may occur even after remission for several years if the patient stops consuming Dd-water.

Cervix

Female patient (47); on Dd-water from 22-11-1993
(Országos Onkológiai Intézet, Budapest)

The patient was diagnosed in November 1993 with an inoperable cervical tumour that was entwined with the ovaries. **Beginning from March 1994, when the patient started to consume Dd-water, her state improved continuously. She gained weight and the tumour in the cervix was no longer detectable. By June the earlier stricture of the intestines also disappeared, the patient was in good general mood.** Starting in November 1994 the patient began to consume water of a less depleted D-content. Results of examinations in February 1995 verified significant progression. We could follow the patient for another five months, during which time she did not consume Dd-water in sufficient quantities, upon which there was a slow regression in her state. This case demonstrates well the connection between the consumption of Dd-water and the behaviour of the tumour, as well as the importance of establishing and maintaining the right dose.

Female patient (56); on Dd-water from 07-12-1992
(SZAOTE Fül-Orr-Gégészeti Klinika, Szeged)

The patient underwent the first surgery in March 1989 with a tumour of the tongue that had existed for a year by then. In 1990, the tumour receded and the patient was given repeated treatment. At the end of 1992, a relapse occurred again, it was then that the patient began to consume Dd-water. From that time on **the size of the tumour decreased, and biopsy performed in March 1993 did not verify tumorous tissue in the sample.** The patient had been consuming Dd-water continuously during the year 1993, and made a break in January 1994. Subsequently there was a significant progression of the illness, and another operation became necessary in August 1994. Dd-water consumption was regular from June 1994 on. In February 1997, another biopsy was performed which again proved to be positive. By increasing the Dd-water dose, the size of the newly appeared nodes decreased, and their substance became softer. A more significant progression was experienced in the spring of 1998 only, when X-ray verified lung metastasis. **Due to the increased dose the patient coughed up slimy secretion, the tumour in her mouth got softer, the size of the tumour near the oesophagus decreased, and the tumour at the ear was no longer sensitive.** The patient survived another year, and died in the spring of 1999. Altogether 10 years had elapsed since the first operation. In the first three years, prior to Dd-water consumption, there were three relapses. We suppose that by a continuous Dd-water consumption, relapses of such an extent would not have happened. The next case serves as a good example for the above.

Female patient (63); on Dd-water from 13-07-1993
(Országos Onkológiai Intézet, Budapest)

The patient had been operated three times before June 1993 with renewed tumours of the oral cavity, then, as the patient refused to consent to the partial elimination of the jaw, she was given a full dose of radiation. At that time she began to drink Dd-water. By August 1993 **the wound under the tongue had healed, the tumour on the chin softened and by September the tumour on the neck disappeared.** From a medical point of view **the years 1994, 1995, 1996, and the first nine months of 1997 went by uneventfully.** Then a wound made by an apricot seed would not heal. The patient was given two cytostatic treatments in November 1997, and the third was missed because the wound had by then healed completely. **The patient, by sustaining her good quality of life is in excellent physical state and has been consuming Dd-water for the sixth year.**

Melanoma malignum

Male patient (51); on Dd-water from 17-11-1994
(SOTE Bőrklinika, Budapest)

In July 1994, the patient had a Clark III melanoma removed. The tumour started from a mole on the left side of the abdomen. In August of the same year block dissection was performed in the left armpit. Following this, the patient was given DTIC and Interferon treatment. He had been consuming Dd-water regularly since the above date and was symptom-free in 1995, 1996, and 1997. During this time a single operation was performed when, **in May 1996, a lymphatic node that had been in the armpit for one and a half years but showed no growth was removed.** The microscopic description of the removed node contains the following: "**... the bulk of the substance is occupied by tumour tissue forming contiguous fields... Frequent dividing formations. Necroses in the substance of the tumour. The tumour is surrounded by a thick fibrous case, the excision does not reveal the extension of the tumour into the adipose tissue.**"

With regard to the long symptom-free period, we made a break in consuming Dd-water after more than 3 years, from the beginning of 1998. One and a half months after the suspension of the consumption of Dd-water, a lump appeared on the breast of the patient, after which he died without having consumed Dd-water, in March 1999.

Male patient (46); on Dd-water from 20-12-1994
(Országos Onkológiai Intézet, Budapest)

The patient had first been operated with melanoma malignum, later, in April 1992, lymphatic nodes were removed from the armpit, followed by further surgery and DTIC treatment in the spring of 1994. By the fall, metastases appeared behind the ear and also in the liver, this was followed by Intron A treatment. **According to the CT scan made one month after the start of the Dd-water cure, two earlier liver metastases could not be verified, two became smaller and one remained unchanged.** Further controls (March, June, December 1996; February, December 1998) described gradual regression of the liver metastases, while, according to the April 1999 CT scan: "**The formerly described small residual lesions are not traceable. Unanimously circumscribed formation is not verifiable by CT scanning.**" **The patient has, since the verification of liver metastases, for the fifth year been active and working.**

Note: We wish to stress that despite the two successful examples mentioned above, melanoma belongs to the type of tumours that react very poorly and can be treated with great difficulty only. We could not achieve any success in cases when week by week newer and newer metastases appeared. This observation of ours is in accordance with the *in vitro* examinations of the American laboratory, according to which the melanoma cell line got adapted to the culture medium of lower D-content within 6 hours, while with prostate and breast cell lines inhibition was traceable for 24–48 hours. Thus, from the aspect of dosage, in the case of this tumour type the dose must be increased, i.e. D-level decreased in the body, within a time shorter than usual.

Malignant tumours of the bone marrow

Male patient (26); AML; on Dd-water from 10-01-1995
(Szt. László Kórház, Budapest)

In September 1992, following a feverish condition, hypertrophy of the lymphoid nodes was reported. With knowledge of the histological diagnosis (Hodgkin lymphoma), first ABVD, then, from February 1994, COPP treatment was given. In June 1994, examinations verified AML M4. After three doses, in October 1994, **the patient refused further chemotherapy. According to blood counts of 3 and 10 January 1995, the number of blastoid cells increased in the peripheries. The patient started Dd-water consumption on 10 January 1995, and the blood count of 31 January could not verify the presence of blastoid cells.** The patient had been drinking Dd-water regularly till March 1997, during this time all the control examinations proved to be negative. From January 1998, for some months, he had been consuming Dd-water as a precaution. **His results have been negative for more than four years.**

Male patient (62); CLL; on Dd-water from 16-01-1996
(Dél-pesti Kórház)

The patient was diagnosed with B-cellular CLL in 1992. **From the end of 1995 on, originating from his basic condition, an increasing leukocyte-count, anaemia, and thrombocytopenia occurred, the patient was treated with concentrates of erythrocytes and thrombocytes.** According to the CT scan, on both apexes of the lung residual fascicles were verified, as well as several lymphatic nodes of 1.5–2 cm in the mediastinum. The liver and the spleen were considerably enlarged; the mesentery contained partly confluent lymphatic nodes. The disease was progressing despite conventional treatment. The patient was weak, mostly bed-ridden, and was constantly losing weight (69 kg). He started consuming Dd-water in January 1996. In the following two months, transfusion was given less frequently, later, on the basis of his blood-count results, the patient did not need transfusion at all. He regained strength, and after four months the lymphatic nodes in the neck were not palpable. By the end of the year, abdominal complaints appeared. In January 1997, several enlarged lymphatic nodes palpable also through the wall of the abdomen were removed surgically. In the spring of 1998, further swollen nodes were found in the armpits. The patient had been drinking Dd-water for 3 months. After one year, **in May 1999, the patient's weight was 80 kg, no swollen lymphatic nodes were palpable, and the patient was in good physical condition.**

Myeloma multiplex

Male patient (65); on Dd-water from 01-11-1994
(SOTE Oktató Kórháza, Budapest)

The patient was diagnosed with myeloma multiplex in September 1994. Prior to this, in March, he had pathologically broken a rib. Cytostatic treatment had been begun two months prior to the start of the Dd-water cure. After two months of consumption, by January 1995, the patient gained 5 kg of weight. In March, electrophoresis verified significant improvement, and after a further two months no pathological deviation was shown. The patient was given the last treatment of the series in July. In September 1996, electrophoresis and blood count were found to be in order; in January 1996 blood levels were normal, and bone scintigraphy could not prove earlier thickenings. Control examinations in September were also negative. In October 1996, the patient took a hot bath cure in Hévíz, by the end of the cure he felt unwell, and had pain between the ribs. In November 1996, a new treatment began. By March 1997, a tumour of a considerable size appeared on the sternum. The tumour was in regression in May, appeared again in August, while metastases appeared between the thoracic vertebrae. The patient was continuously consuming Dd-water, with suspension for some months only. Due to conventional Dd-water treatment, the patient survived a further two years and **died 4 and a half years after the diagnosis.**

Note: Dd-water can, with great efficacy, be used to treat oncological problems connected to hematopoiesis. The consumption of Dd-water significantly enhances the efficacy of conventional treatment and increases the chances to avoid relapse.

Liver

Male patient (66); on Dd-water from 07-11-1994
(SOTE I. sz. Belklinika, Budapest)

A malignant tumour originating in the connective tissue of the liver had been discovered in 1985. Until the start of the Dd-water cure, several operations aimed at decreasing the tumour size were made, and also one closing the blood vessels to block the blood supply of the tumour. In May 1994, because of gastric haemorrhage caused by the breakthrough of the tumour, gastrotomy and an operation to join the stomach with the intestines were performed. Because of jaundice and the fall of blood sugar level, the patient lost consciousness on several occasions. **The patient was hospitalised in October 1994, because of extreme physical deterioration, in a practically hopeless state. Sonography in November recorded a tumour of 17 x 21 cm in size. Following the consumption of Dd-water**, the patient regained appetite, gained 3.5 kg weight and was able to walk. He was emitted from hospital in November. Jaundice decreased, verified also by laboratory results, high enzyme values caused by the tumour fell. **In February 1995, sonography verified a tumour of 16 x 14 cm. One year after the beginning of the Dd-water cure, jaundice almost totally disappeared, and the improvement of gall flow was proved by the fact that the colour of the stool, which had for years been like clay, became normal again.** A significant deterioration of the patient's state occurred in November-December, after which he was unable to regularly consume Dd-water. The patient died on 27 November, 1995. The significant improvement of the patient's state from November 1994 on lasted for a year and can be attributed to Dd-water consumption only, as **the patient was not given any other treatment.**

Neuroepithelioma

Male patient (45); on Dd-water from 25-04-1995
(MÁV Kórház, Budapest)

Severe pain in the back of the patient had appeared in November 1994, upon which L. II–V. compr. fracture was verified but the origin of the primary tumour was impossible to define. CT scan of March 1995 revealed significant progression, which was most expressed at the level of the L. IV. vertebra. As compared to the previous examination, the body of the vertebra was narrower by 0.5 cm. A process of a soft tissue density, destroying nearly the whole body of the vertebra and the right side *proc. transversus* was spreading towards the spinal canal. **Biopsy verified neuroepthelioma. Because of its range and location, the tumour was inoperable. Prior to Dd-water consumption the patient had been bed-ridden. His pain increased somewhat 6-7 days after the beginning of the cure, but 3 weeks later no morphine was needed to be given and the patient was able to get up. Three months later the patient could walk by himself, without using a stick, and was also able to drive a car.** He had been consuming Dd-water till August 1995; no information is at our disposal since that time.

Astrocytoma

Male patient (29); astrocytoma A3; on Dd-water from 28-03-1995
(Országos Idegsebészeti Tudományos Intézet, Budapest)

Following epileptic bouts from 1991, examinations verified the presence of a brain tumour. The operation took place in January 1995. **The patient consumed Dd-water for 44 months without interruption.** During the past years, seizures frequent and strong in the beginning, gradually decreased. **In the past months, several weeks passed without seizures, or, if they happened at all, they were very mild. Then, after 4 years, from March 1999 on, the former 62-ppm preparation was substituted for a 100-ppm one. After one month, strong headache appeared, and after another month a quick deterioration was observed. MRI verified the appearance of a tumour and cyst of the size of a fist.** The increased dose of Dd-water could not inhibit tumour growth. One month later, 4 days after a successful operation the patient was able to leave hospital.

The case is a good example of the significant role of Dd-water in the inhibition of tumour growth. The dose had been appropriate for 42 months, but even after 4 years neither the increase of the D-content of Dd-water, nor the decrease of the dose were justified, as within two months the patient relapsed.

Glioblastoma

Male patient (44); on Dd-water from 05-12-1995
(Országos Idegsebészeti Tudományos Intézet, Budapest)

In August 1995, a left temporoparietal tumour histologically defined as glioblastoma had been removed, after which the patient was given radiotherapy. In January 1996, half a year after surgery, CT scan verified the growth of the ring-like accretion in the operated area. In April 1996, after the patient had been given contrast material, an inhomogeneous contrast accretion was observed with the slight progression of the remaining tumour. In October 1997, CT scan raised the suspicion of a solid recidiva, 1 cm in diameter. In April 1997, the temporoparietal tumour was approx. 3 cm, and the expansive character increased. In 1997, chemotherapy was given. **The patient died after 2 years' survival in November 1997.**

Neurofibromatosis

Female patient (12); on Dd-water from 06-08-1996
(POTE, Onkohaematológia, Pécs)

The young girl had been treated with *opticus glioma*, caused by neuro-fibromatosis, and resulting in amaurosis on both sides. In August 1994, the basic

128

condition caused multiple lesions of the central nervous system, accompanied by decreased hearing, facial, oculomotoric, abducent paresis, and the paralysis of the upper and lower right limbs. **Carboplatin/VP-16 chemotherapy was given,** and, to assess efficacy, MRI was also made. **According to the MRI, a significant expansion occurred, examinations verified progression. Following this, the patient was given no other, conventional treatment. The girl first began to consume, in January 1996, the 130-ppm preparation called Vitaqua available at that time, then, from August, she started to drink an 85-ppm preparation. According to the MRI made in November 1996, the size and contrast matter accumulation of the formerly described focus significantly decreased, the regression of the tumour was verified.** Since the start of Dd-water consumption, the girl's ability to speak and move has continuously been improving, **in January 1997 she was able to walk unaided. In August 1997, contrasted to the state in November 1996, the tumour's size was half of the original.** MRI made in December 1998 showed no progression either. **The girl has been continuing her studies as a private pupil for more than two years now.**

Note: In the case of tumours of the central nervous system, we think that Dd-water can play a role at least as great in the preparation for surgery as later, in post-treatment. With brain surgery, the doctor is often faced with the problem that although he knows that because of the tumour he should remove greater areas, this would affect centres, which endanger the patient's life or causes significant and permanent disability.

Based on our observations so far, it might be advantageous if the patient consumed Dd-water 2–3 months prior to surgery. This would enable the surgeon to remove a more compact tumour isolated from its surroundings to a greater extent. This would also decrease the remaining tumour mass and the area removed would also be smaller. Post-treatment with Dd-water after surgery, in the optimal case, selects remaining tumour cells for good.

Further positive experiences were gained with the following tumour types:
      chronic myeloid leukaemia (CML)
      acute lymphoid leukaemia (ALL)
      neuroblastoma
      cancer of the testicles
      cancer of the bladder
      cancer of the kidney
      Non-Hodgkin lymphoma
      Hodgkin lymphoma
      thyroid cancer.

MAIN RULES OF APPLICATION (I–IX)

We have defined the main rules of the application of Dd-water as follows.

I

The dose applied should be defined in harmony with the patient's condition. The aim is to ensure the minimal effective dose and to maintain the decrease in D-concentration as long as possible.

Based on our experiences, tumour regression occurs gradually (in weeks, months or years). We suppose that the reason for this is that the cell is sensitive to D-depletion in a given phase of its cycle only. When the decrease in D-concentration occurs, it is only cells at a given point of the cell cycle that necrose. Cells, however, which are not in a phase sensitive to D-depletion, get adapted to the slowly decreasing D-concentration. If, by dosage, we lengthen the reducing process of D-concentration as required, we can achieve that in the body there will be a decrease in D-concentration even when cells, which have adapted to a lower D-concentration, enter the given phase of cell division. Suddenly administered great doses may produce quick results at the beginning of the treatment, but later, if further decreases of D-concentration are not possible, cells in a resistant phase, having adapted to Dd-water medium, will be able to proliferate. The quick decrease of D-concentration may be justified with sensitive tumour types, or where tumour mass is not significant (1–2 cm in diameter), but when tumour mass is considerable (e.g. colon tumour with a multiple expansion in the liver), according to our present knowledge, a lower initial dose and its slow increase may bring better results.

II

It is not advised to suspend Dd-water consumption while the tumour is detectable in the body.

Dd-water consumption exerts a selective stress on cells, to which tumorous cells are more sensitive than healthy ones. During the application of Dd-water, in the beginning of the treatment we decrease D-concentration and later we ensure a constant D-concentration in the patient's body, on a level significantly lower than natural concentration. The cessation of this selective stress, i.e. the suspension of Dd-water consumption, results in an increasing D-concentration, which, according to our experiences so far, may lead to tumour growth. The process, however, seems to be reversible, which means that the resumed consumption of Dd-water may again stop tumour growth, but then the result achieved at the first application can no longer be reached. This observation justifies that the patient should suspend Dd-water consumption in medically reasonable cases and after relevant control examinations only.

130

III

Dd-water cure should be continued even after complete remission, for a minimum of 6–10 months. The duration of post-treatment may vary according to the type of the tumour (e.g. with tumours of the ovaries, breast, glioblastoma, astrocytoma, etc., a longer post-treatment may be needed).

Our general experience is that the patient should feel threatened only if he arbitrarily stops treatment before time. Following from the character of the disease, the most difficult thing to decide is when the patient can be considered as healed. In this respect, it is of special significance that the treatment of the first patient began on 30 October 1992, and since that time we have followed several hundred patients for several years. (The first patient began consuming Dd-water with inoperable prostate cancer, urination problems and a slightly elevated PSA value. His urination problems stopped within two weeks, PSA value also decreased, and after a month the tumour was no longer palpable. The patient had been consuming Dd-water for a year, and has been free of symptoms and complaints ever since.) During the past years we could successfully suspend the treatment with several patients without relapse, but unfortunately, it also happened that a patient had a relapse within months following the suspension of Dd-water.

IV

Before suspending or finishing the Dd-water cure, the dose should be decreased gradually.

The sudden cessation of Dd-water consumption results in the quick increase in D-concentration in the body. In the case of patients who had stopped Dd-water consumption before time, a relatively quick deterioration was observed. This may be explained by the fact that the increase of D-concentration creates favourable conditions for tumour cells to proliferate. To patients who stop Dd-water consumption in a controlled manner, or to those with whom we suspend consumption because they have for a longer time been in remission, a lengthy and slow decrease of the dose is recommended. (We presume that D-concentration thus slowly nearing the original cannot be used by eventually still present tumour cells to such an extent as if original D-concentration returned suddenly.)

We consider it to be of special importance that Dd-water should be withdrawn gradually, if the patient had, for a longer time, been consuming Dd-water of 30–50 ppm D-content.

V

With patients in remission, after a break of 2–3 months we recommend to resume the cure even if the patient has no complaints at all. In such cases, the renewed 4–6 months consumption of Dd-water may be followed by a longer break of 4–6 months, followed by a new cure lasting for 2–3 months.

The repeated consumption of Dd-water is justified by the fact that we cannot state that after a symptom-free period of 1 or 2 years, we have really succeeded in selecting all initial tumours. The repeated consumption of Dd-water inhibits cell division and may destroy possibly existent cell groups, which, in time, may result in macroscopically demonstrable tumours. We have several experiences when no tumour could be verified by imaging but the change in tumour marker values demonstrated the effect of Dd-water well (they decreased after the consumption of Dd-water).

VI

It is not recommended to decrease the daily dose during treatment, only to increase it or keep it at a constant level.

The grounds for this is, that it should never happen even with daily fluctuation that D-concentration temporarily increases in the body. Experience tells us that patients who consume their doses during the day in even amounts, by 2–3 dl, show better results than those who consume the total amount on one or two occasions.

VII

It is advisable to ensure the daily fluid intake in the form of Dd-water.

The optimal decrease of D-concentration may be reached if the patient, within rational limits, consumes Dd-water only. This does not exclude the preparation of soup or other meals with normal water, but drinking water should be ensured from Dd-water. The establishment of the daily dose must, thus, be harmonised with the fluid needs of the patient.

VIII

The increase of concentration resulting from normal water and food consumption should be balanced by the consumption of small volumes (0.5–1 dl) of 25-ppm Dd-water.

Experience gained in the past years has unanimously verified that the increase of concentration resulting from untimely suspension of Dd-water consumption favours tumour cell growth – the tumour is able to grow again, and this fact is verifiable with diagnostic instruments within weeks. We suppose that it is also favourable for tumour cells if, after food intake, selective stress decreases for a while, because the normal deuterium content of the food temporarily increases the D-concentration in the body. To counter-balance this we recommend the consumption of 0.5–1 dl undiluted 25-ppm Dd-water after meals.

IX

The doctor should check changes and follow the effect of Dd-water at first 6–7 days then 18–21 days after beginning the Dd-water cure, later once a month.

In the first few days usually no significant changes occur, but as we are dealing with the application of a new preparation, the close follow-up of the patient is justified. Eighteen to twenty-one days after beginning the Dd-water cure the patient usually becomes spiritless and somnolent. Further monthly control is justified by the continuous and timely correction of the dose of Dd-water.

GUIDE AND TABLES TO DOSAGE

Dosage schemes, in the case of detectable tumours, are based on direct observation. With patients in remission as well as the healthy population we also have direct experience but our recommended dosage has been extrapolated using the data of cancer patients in the first place. (We suppose that if within a time unit cancer patients show significant regression, then with patients in remission initial tumours below the detectable level also become necrotic on the effect of Dd-water.)

We have divided the population into two main groups; the guide to preventive dosage for the healthy population (Group H) is dealt with in *Chapter I* of *Book One.* In the chapter written for doctors we have summed up our recommendations for patients diagnosed with cancer and already treated (Group C). This group is further divided (Groups C/R and C/T). To Group C/R belong patients who had previously been diagnosed with tumour but presently no tumour can be verified (total remission R), while Group T are those, in whom the tumour (T) is still present. Both groups (C/R and C/T) are further subdivided.

Concentration values in the protocol should be ensured for the patient either directly by administering Dd-water of the given concentration, or by mixing water of a low D-content with natural water according to certain rates (*Table II. 7*). From water of a low D-content, a preparation of a concentration lower than 15 ppm but higher than the original can be produced. With concentration values the mean value is given, a deviation of ± 5 ppm is within the limit of tolerance. (It should be noted that in protocols the lowest D-concentration is 25 ppm. It is possible to produce water of a lower D-content but presently it is not available.)

GROUP C

CANCER PATIENTS IN REMISSION (GROUP C/R)

Patients who had already been diagnosed with a malignant tumour of some kind but at present are free of symptoms constitute this group. These patients get therapy (surgery, chemotherapy, and radiation therapy) and thanks to this, they are free of tumours. The differentiation of two sub-groups is justified:

C/R–I

The patient subsequent to the first treatment or series of treatments had a remission, and treatment was finished more than one year ago. There has been no relapse since the set-up of the diagnosis and treatment, the patient is free of symptoms and complaints; the latest control examinations are not older than 3–4 months and are negative.

Within this group further sub-groups could be made according to whether the chances are greater for healing or a relapse. Since this is almost impossible to decide in the case of a given patient, the dosage protocol was elaborated to avoid occurrence of the worst event, i.e. relapse. The application of this might seem overinsured in the case of a patient with good prospects but it also increases the chances for healing.

For patients in the C/R–I group we advise the consumption of Dd-water according to *Table II. 14*. It is recommended to repeat the protocol after a break of 1–2 months. Following this, H-protocol should be repeated yearly, for a period of 3–4 years.

Table II. 14 *Dosage guide for patients belonging to Group C/R–I*

Recommended D-concentration	Recommended daily dose	Recommended duration of treatment
87.5 ppm	1.0–1.4 litres	5–6 weeks
75 ppm	1.0–1.4 litres	7–8 weeks
62.5 ppm	1.0–1.4 litres	10–12 weeks
50 ppm	1.0–1.4 litres	10–12 weeks
37.5 ppm	1.0–1.4 litres	6–7 weeks
50 ppm	1.0–1.4 litres	3–4 weeks
62.5 ppm	1.0–1.4 litres	3 weeks
75 ppm	1.0–1.4 litres	2 weeks
87.5 ppm	1.0–1.4 litres	1 weeks
100 ppm	1.0–1.4 litres	1 weeks
112.5 ppm	1.0–1.4 litres	1 weeks
112.5 ppm	0.5–0.7 litres	1 weeks

C/R–II

The patient is tumour-free and post-treatment has been finished within one year or is still going on. This includes patients in remission, who underwent at least one relapse but due to renewed treatment, remission was achieved again. In this group, despite remission, relapse may be expected, as there had been, e.g. distant metastases; the size of the primary tumour had been more than 1 cm; or the tumour type is characterised by the appearance of metastases, etc.

For the treatment of patients belonging to type C/R–II, we recommend consumption of Dd-water according to data in *Table II. 15*. It is advisable to repeat the cure after a break of 2–3 months at least once. This should be followed by protocol C/R–I, taking the shorter recommended duration (e.g. 5 weeks, 7 weeks, 10 weeks, etc.) and protocol H yearly, for a period of 2–3 years.

Table II. 15 *Dosage guide for patients belonging to Group C/R–II*

Recommended D-concentration	Recommended daily dose	Recommended duration of treatment
87.5 ppm	1.0–1.4 litres	5–6 weeks
75 ppm	1.0–1.4 litres	7–8 weeks
62.5 ppm	1.0–1.4 litres	10–12 weeks
50 ppm	1.0–1.4 litres	14–16 weeks
37.5 ppm	1.0–1.4 litres	14–16 weeks
25 ppm	1.0–1.4 litres	5–6 weeks
37.5 ppm	1.0–1.4 litres	2–3 weeks
50 ppm	1.0–1.4 litres	2 weeks
62.5 ppm	1.0–1.4 litres	2 weeks
75 ppm	1.0–1.4 litres	1 weeks
87.5 ppm	1.0–1.4 litres	1 weeks
100 ppm	1.0–1.4 litres	1 weeks
112.5 ppm	1.0–1.4 litres	1 weeks
112.5 ppm	0.5–0.7 litres	1 weeks

PATIENT POPULATION WITH DETECTABLE TUMOUR (GROUP C/T)

For cancer patients we have set up the following three categories.

C/T–E (early stage)

The illness had been diagnosed, treatment was or is going on but complete remission could not be achieved or, if so, the patient relapsed; or conventional treatment is expected to result in the patient's remission but this can be maintained for a short period (3–6 months) only. Life expectancy of the patient considerably surpasses one year.

C/T–P (progression)

Tumour or tumours are detectable in the patient; treatment may have been going on for months. Temporary improvement or stagnation may be achieved but the condition, as a whole, is progressing, no complete healing is expected. Despite the above, the patient is in good physical condition (ECOG: 0–2) and life expectancy surpasses half a year.

C/T–A (advanced)

Irrespective of the date of the diagnosis, this category refers to patients who have got under medical control at an advanced stage; the condition has been treated for years but slowly all possibilities are exhausted; because of the character of the condition (tumour of the pancreas, disseminated melanoma, glioblastoma, etc.) life expectancy does not exceed, or only slightly surpasses half a year.

PROTOCOLS FOR PATIENTS IN GROUP C/T

It can generally be stated that, according to our present knowledge, those belonging to group C/T–E should consume the preparation for a minimum of 2–3 years; patients in Group C/T–P for 3–4 years; and those in Group C/T–A, provided they succeed in surviving the critical phase, should consume Dd-water for up to 4–5 years (eventually with shorter breaks). If the patient does not succeed in getting into remission, we do not recommend the suspension of Dd-water consumption, or only after a 4–5-year-long consumption for 1–2–4 months. When establishing dosage strategy, the pace of D-depletion should be planned according to the above periods. Establishing the duration of the treatment depends, in the first place, on when remission can be achieved. **The sooner complete remission can be achieved and the longer the patient consumes Dd-water, the greater the probability of avoiding relapse.** With patients where no remission was achieved, the application of Dd-water slows down tumour growth, and may result in partial regression, and may increase the efficacy of conventional treatment and survival chances. When applying Dd-water, it should be kept in mind that improvement or healing may be traced by two main factors:

136

a) if the D-concentration in the patient's body continuously decreases, or
b) if a D-concentration lower than natural is continuously maintained in the body.

With the population belonging to group C/T, where, at the beginning of Dd-water treatment the tumour is always detectable, treatment, in an ideal case, consists of two phases:

a) achieving complete remission;
b) maintaining remission.

Remission can be achieved by decreasing the D-concentration of the body. The decrease of D-concentration can be performed according to three main strategies of dosage:

– gradual (C/T-g) protocol;
– fast (C/T-f) protocol;
– slow (C/T-s) protocol.

If any of the above protocols have been carried out and the patient arrives at consuming water with the lowest D-content, the following may be experienced:

a) the patient has reached remission during the cure;
b) the protocol is over but the patient did not reach remission.

a) If the patient has reached remission, it is, of course, not indifferent whether this happened at the beginning of the treatment or towards the end of it. In the first case the patient has been consuming Dd-water while being in remission, whereas in the second case this situation has existed for some months only. It can generally be stated that the patient should consume Dd-water for a minimum of 8–10 months even after reaching complete remission. We consider it to be of utmost importance to enhance the range of experiences, as the shortness of time and the low number of cases does not allow us to state exactly after how many months of tumour-free period can a patient be considered as healed or healthy. When dealing with the main rules of application, we mentioned that with some tumour types an 8–10-months tumour-free period does not necessarily mean complete healing. If the patient continues the consumption of Dd-water beyond the recommended 8–10 months, he further improves his chances of healing. If the doctor believes that, because of the long lasting remission, the selective stress maintained by Dd-water on supposed tumour cells can be decreased, the patient can be considered to belong to Group C/R instead of Group C/T. Transition to stopping or interrupting the cure is recommended according to protocol C/R-end. If the protocol ends, treatment should be carried on according to the summary of the main rules, but following the C/R–I or C/R–II protocol.

b) If the patient had finished the protocol but could not reach remission, it is advisable to continue the protocol by consuming water of the lowest D-content.

Based on achievements so far, we may state that with the majority of patients belonging to Group C/T–E, remission occurs within 1–1.5 years, in some cases even within some months.

The preparation can be applied successfully with the majority of patients belonging to group C/T–P. It is not easy to define results that can be achieved routinely, but in this group, following Dd-water consumption, regression may occur even if, despite conventional treatment, the condition had earlier been progressing. Depending on the size and type of the tumour, the patient may slowly become tumour-free. It is an important aspect that in case the illness should progress again after years, the patient may return to cytostatic drugs to which the tumour had earlier been resistant.

With patients belonging to Group C/T–A, remission can also be reached. Success may be achieved with various types of leukaemia, even in an advanced state, but D-depletion has proved to be effective with several other tumour types (e.g. that of the breast) as well. Results are difficult to achieve in the case of solid tumours of a great mass, e.g. those filling the lesser pelvis, especially if they are of a colorectal origin. Based on our present knowledge, in the majority of cases, with patients belonging to Group C/T–A the realistic aim is to improve the quality of life of the patient by causing regression and to inhibit tumour growth for up to several years. If the patient is over the first several months of Dd-water consumption, in some cases unexpected improvement may occur even after years. In these cases, we suppose that the tumour "gets exhausted" in a medium of low D-content. It is a general experience that the patient should have at least 3 months of life prospects for the positive effect of Dd-water to assert itself in the long run.

PROTOCOLS C/T

As to the dosage of Dd-water for patients in Group C/T, we recommend three different schemes, and one for the termination of the treatment. We consider it important to stress that our recommendations serve as a basis for beginning the treatment. Elaboration of exact protocols, in the light of a great number of parameters, is a task to be solved in the future.

PROTOCOL C/T-G

D-concentration is constantly decreasing in the patient's body.

This protocol is recommended in the case of patients belonging to Group C/T–E. If no relapse occurred from the beginning of Dd-water consumption, after a **minimum** of 8–10 months' symptom-free period, a short, maximum 2-months long break may be held. After this the treatment should continue with protocol C/R–I or the administration of 25-ppm water maintained for a further half year.

Table II. 16 *Dosage guide primarily for patients belonging to Group C/T–E*

Recommended D-concentration	Recommended daily dose	Recommended duration of treatment
87.5 ppm	1.0–1.4 litres	6–8 weeks
75 ppm	1.0–1.4 litres	6–8 weeks
62.5 ppm	1.0–1.4 litres	8–10 weeks
50 ppm	1.0–1.4 litres	8–10 weeks
37.5 ppm	1.0–1.4 litres	8–10 weeks
25 ppm	1.0–1.4 litres	min. 8–10 weeks

PROTOCOL C/T-S

D-concentration slowly decreases in the patient's body.

This protocol is recommended primarily for patients belonging to Group C/T–P. If the total mass of the tumour is considerable, (primary tumour of 2–3 cm in diameter, numerous distant metastases, etc.) treatments may start with 75-ppm Dd-water. Deviation from the above dosage scheme, inasmuch as it means consumption for a longer time or the decrease of D-concentration, further increases the probability of healing or maintaining remission.

Table II. 17 *Dosage guide primarily for patients belonging to Group C/T–P*

Recommended D-concentration	Recommended daily dose	Recommended duration of treatment
87.5 ppm	1.0–1.4 litres	6–8 weeks
75 ppm	1.0–1.4 litres	8–10 weeks
62.5 ppm	1.0–1.4 litres	12–14 weeks
50 ppm	1.0–1.4 litres	12–14 weeks
37.5 ppm	1.0–1.4 litres	14–18 weeks
50 ppm	1.0–1.4 litres	14–18 weeks
25 ppm	1.0–1.4 litres	min. 18–20 weeks

PROTOCOL C/T–F

D-concentration rapidly decreasing in the patient's body.

This dosage strategy is recommended for patients with whom the tumour is aggressive; the prognosis is poor, and life expectancy is about half a year (C/T–A). Such are melanoma, tumour of the pancreas, acute forms of leukaemia or sarcomatoid tumours; or if several relapses had occurred beside conventional treatment. If the protocol has ended, it is advisable to maintain low D-concentration by administering 25-ppm Dd-water.

Table II. 18 *Dosage guide primarily for patients belonging to Group C/T–A*

Recommended D-concentration	Recommended daily dose	Recommended duration of treatment
87.5 ppm	1.0–1.4 litres	2 weeks
75 ppm	1.0–1.4 litres	2–3 weeks
62.5 ppm	1.0–1.4 litres	3–4 weeks
50 ppm	1.0–1.4 litres	5–8 weeks
37.5 ppm	1.0–1.4 litres	8–10 weeks
25 ppm	1.0–1.4 litres	min. 10–12 weeks

PROTOCOL C/R–END

The gradual termination of the treatment.

If the patient is symptom-free, by slowly increasing D-concentration, having reached the normal level, a break of 1–3 months is recommended.

Table II. 19 *Dosage guide before the suspension of Dd-water administration*

Recommended D-concentration	Recommended daily dose	Recommended duration of treatment
37.5 ppm	1.0–1.4 litres	1–2 weeks
50 ppm	1.0–1.4 litres	1–2 weeks
62.5 ppm	1.0–1.4 litres	1–2 weeks
75 ppm	1.0–1.4 litres	1–2 weeks
87.5 ppm	1.0–1.4 litres	1 week
100 ppm	1.0–1.4 litres	1 week
112.5 ppm	1.0–1.4 litres	1 week
112.5 ppm	0.5–0.7 liters	1 week

PROTOCOL H

When establishing dosage, we supposed an average body mass of 55–70 kg. The daily dose of 0.5–0.7 litres is considered to cover half of the average daily water intake. (We note that from 105-ppm water, mixed in a fifty-fifty ratio, we can get 127.5-ppm water. The cure can also be continued so that a daily 1–1.7 litres of Dd-water is consumed in the first and last 2–3 weeks of the cure, in the form of a double solution of 105-ppm water.)

Table II. 20 *Dosage guide for healthy people and cancer patients in remission for 2–3 years*

Recommended D-concentration	Recommended daily dose	Recommended duration of treatment
105 ppm	0.5–0.7 liters	the first 2–3 weeks
105 ppm	1.0–1.4 liters	the second 2–3 weeks
105 ppm	0.5–0.7 liters	the third 2–3 weeks

DEVIATIONS FROM THE ABOVE PROTOCOL

The type and location of the tumour, the body mass of the patient and numerous other parameters may influence the establishment of dosage, as detailed in *Chapter III*. We only mention some of these:

1. In the case of certain tumour types Dd-water may get in direct contact with the tumour. With these tumour types (those of the oral cavity, oesophagus, larynx, stomach, etc.) treatment may begin with a water of a higher (100–112.6 ppm) D-content.
2. Volumes stated in the protocol may, naturally, be decreased if, e.g. children (5–40 kg) or emaciated adults consume Dd-water; and should be increased if an adult of 90–100 kg or more is being treated.
3. With patients of a greater body mass or belonging to Group C/T–A, treatment may, or should begin with water of a lower (75.0, 62.5 ppm) D-content.

FURTHER OBSERVATIONS

THE EFFECT OF Dd-WATER ON *DIABETES MELLITUS*

In recent years, Dd-water has been examined exclusively from the aspect of its anti-cancer effect. We have had, however, some patients with tumours and diabetes mellitus at the same time. It attracted our attention that after drinking Dd-water, the state of *diabetes mellitus* improved with these patients. This manifested itself by the fact that their blood sugar levels that had earlier been fluctuating, became steady, or high blood sugar levels decreased, more or less to the normal range. We started to follow the level of blood sugar with patients suffering from *diabetes mellitus* and we could definitely state that Dd-water positively influenced the process of *diabetes mellitus*. On the basis of the results so far, the positive effect of Dd-water seems to have been demonstrated both with Type I and II *diabetes mellitus* (DM). With patients suffering from diabetes mellitus who drink Dd-water, especially those belonging to Type I (insulin-dependent patients) the changes in blood sugar level should be followed. With these patients the level of blood sugar will decrease earlier, which can be compensated for a while with intake of sugar, but it is preferable to keep blood sugar level in balance with the decrease of the insulin dose. With patients suffering from Type II *diabetes mellitus*, the dose of the medication should also be decreased in some cases but this group seems to be less sensitive. This means that the absence of the dose correction of the medication does not have consequences of the same extent as with patients of the insulin-dependent group, if insulin dosage is not corrected. It should be emphasised that the above positive effects occur in some weeks or months, parallel with Dd-water consumption, therefore unexpected, acute situations will only arise if the doctor does not follow and correct carefully the slowly changing blood sugar level.

THE POSITIVE EFFECT OF Dd-WATER ON MYOMAS, CYSTS AND BENIGN HYPERTROPHIES OF THE PROSTATE

Thanks to the great number of patient population, we have met numerous cystic deformations. It can generally be stated that cysts proved to be sensitive to Dd-water; mainly those of a smaller size had completely disappeared within a short time. It was only a few patients who consumed Dd-water because of benign myomas or hypertrophies of the prostate but in these cases we also experienced improvement.

SIGNIFICANT WEIGHT GAIN

From a thousand patients consuming Dd-water, in 4–5 cases we experienced significant weight gain (10–20 kg). In one case examination verified the malfunctioning of the thyroid gland, which was treated with medication. At present we cannot decide yet whether there is a cause and effect connection between the consumption of Dd-water and weight gain.

FURTHER TASKS

We referred to the contradictions of the development of anti-cancer drugs by comparing it to flying. To further expand this comparison, we might say that we have found a principle to construct a "flying machine". The real challenge here is to construct machines that can fly very far with great security. As the making of the first flying machine implied the possibility of giant Boeings carrying hundreds of passengers overseas, the process based on deuterium depletion also implies that doctors may cure millions of cancer patients routinely, as it is done with tuberculosis and other diseases. This can only be realised if intense research and development begins in this field. Just as the development of the Boeings was enabled by a series of discoveries and developments in fields of science sometimes distant from each other (informatics, rubber industry, aerodynamics, space research, physics of solid bodies, etc.), in our case tumour medicine will also be advanced by discoveries, recognitions, and unimaginable developments. The main task for doctors is to discover, step by step, during carefully planned clinical trials, the principles and rules referring to the dosage and application of Dd-water. They should find out, from among the presently used drugs and processes which ones increase the effect of Dd-water and which are the most favourable combinations of available possibilities for the patient. If we fulfil our task well, results and experiences will quickly increase, and this may improve the efficacy of healing work day by day.

Colleagues at HYD Ltd. for Research and Development are doing their best to co-ordinate, promote and organise research and development of deuterium depletion, and they are constantly sharing collected information with scientists and physicians all over the world.

PREPARATIONS BASED ON Dd-WATER

During the seven years of the approval process, the funds needed for development meant the greatest difficulty. The costs of the development of a drug may go up to 400–500 million dollars, which is 100–125 billion forints. Because of the high costs, development remained the privilege of some multinational giant companies but even these frequently merge to decrease per unit costs of development.

With knowledge of the above it seems ridiculous that in Hungary, a company founded with a capital of 1.5 million forints (not dollars!) should start a process of approval. The reason for doing so might be due to two causes: first, we believed that drug development based on deuterium depletion may open new vistas in drug research; and second, that we had no other choice, because the idea seemed to be so unbelievable that there was nobody to take it and the inherent possibilities seriously.

To bridge constant financial problems, we tried to develop products which are connected to our original idea, deuterium depletion.

Our first such product was the deuterium-depleted drinking water named **Vitaqua.** This had been approved as a food product in which D-concentration was decreased, compared to normal drinking water, only slightly (by 20 ppm).

Our second product was the **YUVAN** family of cosmetics developed jointly with Florin Ltd. This contains active material of depleted deuterium content under the name Hydro Light. High standard quality and the appearance of the family of ointments was approved by both specialists and consumers but because of the narrow marketing frames the message of the product could not reach consumers. After the first great upswing, sales dropped. The family of products can be ordered through the parcel service of Florin Ltd. (Szeged), and the body lotion is sold as a product of TESCO.

Our third product was approved in the summer of 1999. Under the name **VETERA-DDW-25**, the first anti-cancer drug for veterinary purposes, based on deuterium depletion, was born. The treatment of tumorous cats and dogs became possible and it is to be expected that veterinary surgeons applying VETERA-DDW-25 will be convinced, within a short time, that an anti-cancer drug more effective than any other so far is at their disposal.

Our fourth approved product is **Preventa-105**. The product was marketed in the summer of 2000 as a drinking water. As is clear from the name, the D-concentration of Preventa is 105 ppm, 45 ppm lower than that of normal drinking water.

The approval process of the anti-cancer human drug registered under the name **Depletin** and based on D-depletion is currently under way.

DOSAGE RECOMMENDATIONS AND DIRECTIONS OF USE FOR VETERINARY SURGEONS

APPLICATION OF VETERA-DDW-25

In the following we present the dosage scheme accepted during the approval process. *Table A1* reveals that the principle of dosage is identical with that of human application. The aim of the treatment is to decrease the deuterium concentration in the sick animal's body by applying Dd-water in such quantities and quality (relevant D-concentration) that it should reach the maximum effect. When establishing dosage, we recommend considering the aspects given in the chapter on human application. As the deuterium content of VETERA-DDW-25 is low (25 ppm), the preparation should be diluted according to the table below:

Table A1 *Changes in the dosage of VETERA-DDW-25 as a function of the duration of treatment*

Duration of treatment	Mixing rates of VETERA-DDW-25 and drinking water	Final D-concentration
months 1–2	0.5 l VETERA-DDW-25 + 0.5 l drinking water	87.5 ± 10 ppm
months 3–4	0.6 l VETERA-DDW-25 + 0.4 l drinking water	75.0 ± 10 ppm
months 5–6	0.7 l VETERA-DDW-25 + 0.3 l drinking water	62.5 ± 10 ppm
months 7–10	0.8 l VETERA-DDW-25 + 0.2 l drinking water	50.0 ± 10 ppm
months 11–14	0.9 l VETERA-DDW-25 + 0.1 l drinking water	37.5 ± 10 ppm

Table A2 *Recommended daily dose of the animal as a function of body mass*

Body mass	Volume of Dd-water
1–5 kg	25–75 cm^3
6–10 kg	75–150 cm^3
11–15 kg	150–300 cm^3
16–20 kg	300–400 cm^3
21–30 kg	400–500 cm^3
31–40 kg	500–600 cm^3
41–50 kg	600–750 cm^3
51–60 kg	750–1000 cm^3
Above 61 kg	1000–1500 cm^3

From Dd-water of the suitable D-content, daily dose should be established according to *Table A2*.

It is advisable to divide the daily dose into 2–3 parts. The given amount should be drunk within some hours after pouring it out. Contact with air for a long time unfavourably influences the composition of the preparation, as D-concentration increases. When once 90-ppm water was left in an open glass vessel in the lab, after 48 hours it showed a 10-ppm increase in D-concentration. It follows from the above that the 150-ppm vapour in the air increases the D-concentration of the water, but it also shows that the process is rather slow, so contact with open air does not cause significant changes in the composition of Dd-water.

Contraindications

During research with Dd-water we carried out teratological experiments on pregnant rats. We have not experienced any disorder when examining the offspring but the number of cases did not reach the minimal value needed for statistical evaluation. As it cannot be stated with absolute certainty that Dd-water has no effect on pregnant animals, we do not recommend its consumption in pregnancy.

Side-effects

The most common side-effects are despondency, weakness, and somnolence. With animals these symptoms generally appear some weeks after the beginning of the treatment and last for some weeks. The phenomenon is attributed to tumour necrosis evoked by Dd-water and indicates its effect. From the point of view of dosage, the follow-up of accompanying phenomena is of great importance, because we only recommend the change of the mixing ratio when despondency and somnolence accompanying the actual dosage have already stopped. It should be noted that these side-effects usually appear if the tumour mass is relatively great and if the reaction evoked by Dd-water is quick. This is the case with tumour types sensitive to D-depletion. If the animal undergoes surgery without any greater amputation, operation is advised. If the animal is inoperable, it might become operable on the effect of Dd-water consumption.

Cytostatic treatment in general is not justified in case of animals, because neither its efficacy nor its high costs make it an alternative for Dd-water treatment. Animals drink Dd-water with pleasure and tolerate treatment well, while toxic chemotherapy causes death in many cases.

Further tasks

Research into the effect of deuterium depletion is motivated by the aim to find a new and effective way of treating tumorous diseases. As with every drug development, the next step after the development of the active material was the establishment of the optimal dosage. After the first successful experiments with mice, it became evident that mice with an average body mass of 25–30 grams are not suitable for providing data to extrapolate for humans 3 500 times their body mass. This is why our attention turned to cats and dogs, because their body mass (4–45 kg) may be compared to that of humans, and daily Dd-water consumption can be exactly followed to establish an optimal dosage. Statements made in connection with cancerous cats and dogs may and should be used during application for humans. The work and observations of veterinary surgeons have significance in human medicine.

We have to carry on working in order to establish the appropriate dosage scheme according to the sensitivity of different tumours. Animals should be followed for a long time and a post-treatment strategy should be elaborated to achieve complete healing and to prevent relapses.

It is important to find out the connection between the following: body mass – tumour mass – the dose of Dd-water – D-concentration; as well as to find out which parameters to change die demonstrable from the blood, and which may be followed during the application of Dd-water.

HYD Ltd. for Research and Development co-ordinates research connected to the application of Dd-water and is ready to meet interested veterinary surgeons.

ACKNOWLEDGEMENTS

If you want to achieve something, the whole Universe joins forces to help you.

If deuterium depletion will finally occupy its place in cancer therapy, this will be due to the faith, work, endurance, support and goodwill of many people. Without them application of Dd-water for medical purposes would have remained an idea only. I wish the history of Dd-water would serve as a proof that a good idea can be realised in Hungary as well; one does not necessarily have to "go West". When I express my gratitude, I also speak for all of us who are thankful to those who have helped us in our work, because their assistance and enthusiasm may render our lives better and happier.

Let me express my gratitude first of all to my wife who, not only as a biologist, but as a companion helped my work and many a time, as here, in this book too, gave a final form to my thoughts.

Thanks are due to those members of the Hungarian scientific community whose curiosity and thirst for knowledge urged us to get proof of the biological effect of deuterium depletion in the only possible way: by experimenting. I would like to mention Gábor Jancsó Ph. D. and György Jákli Ph. D. first of all, who, as researchers at the Atomic Energy Institute, were among the first to provide help and secure professional background for our work.

We are grateful to researchers in Szeged: Gábor Laskay Ph. D., Zoltán Galbács Ph. D., Sándor A. Kiss Ph. D. and Gábor Galbács Ph. D.; SOTE (Semmelweis Medical University) researcher Miklós Molnár Ph. D. and veterinary surgeon Tamás Berkényi Ph. D. at the Alpha-Vet Veterinary Hospital.

We would like to express our special thanks for their co-operation in Phase II clinical trials to the following hospitals and doctors: Szt. János Kórház, Department of Urology headed by Professor Géza Wabrosch and György Szolnoki M. D.; Uzsoki Kórház, Department of Urology headed by Professor Barnabás Ruszinkó; Flór Ferenc Kórház Department of Urology under Endre Szüle M. D. and Szt. István Kórház, Department of Urology headed by Dénes Répássy M. D.

Thanks are due to the Departments of Oncology of the following hospitals for their co-operation in the trials on the basis of compassionate use: Heim Pál Gyermekkórház (Budapest); Szt. Borbála Kórház (Tatabánya); Szt. László Kórház (Budapest); Szt. Ferenc Kórház (Miskolc); Markóth Ferenc Kórház (Eger). Beyond the above we are especially grateful to Zoltán Zombori M. D. (Flór Ferenc Kórház, Department of Oncology, led by Kinga Kammerer M. D.) and Zsuzsanna Hódi M. D. (Bács-Kiskun Megyei Kórház Centre for Oncoradiology led by Miklós Szűcs M. D.).

Institutes, which joined our work for a shorter or longer period during the past nine years, also deserve our gratitude: Országos Onkológiai Intézet; Budapesti Műszaki Egyetem; SOTE 1. sz. Pathológiai és Kísérleti Rákkutató Intézet; SOTE Kémiai Biokémiai Intézet; Kertészeti és Élelmiszeripari Egyetem; MTA Növényvédelmi Kutató Intézete; MTA Szegedi Biológiai Központ; JATE Szervetlen és Analitikai Tanszék; JATE Növénytani Tanszék; Keszthelyi Toxikológiai Intézet; Országos Közegészségügyi Intézet.

We were helped in our work by researchers and doctors in Switzerland, Germany, Canada, Great Britain and the United States of America. We would like to say special thanks to John P. Fruehauf and Ricardo Parker who, as medical and research directors of Oncotech Inc. have corroborated our findings by several independent examinations.

Mention must be made of the support of the staff at Buszesz Co., which enabled us to produce and bottle Dd-water on a level appropriate for the requirements.

The head and staff at Reanal Finomvegyszergyár Co. have helped our work by establishing an adequate background of drug production and quality assurance.

We would like to stress the open and innovative approach and professionalism of the Florin Co. at Szeged; they developed the YUVAN family of cosmetics of an excellent quality and compound within months.

Thanks are due to the Danubia Szabadalmi és Védjegy Iroda Ltd., especially to our patent attorney Dr. Tivadar Palágyi and his colleague Márta Király for the high level representation of our patents.

Grateful acknowledgements are due to writers and editors who followed our work with attention and enabled the information to reach the general public. We would like to stress the staunch attention of writers Kata Réz and József Hoffman who have published several articles connected with our subject, thus promoting the approval process of Dd-water. Without aiming at completeness we would like to mention the following journals, papers and TV-staffs: *Magyar Nemzet* (Ildikó Hankó); *Népszava* (András Lantai); *Heti Világgazdaság* (Tamás Vajna); *Mai Nap* (Kata Réz); *Délmagyarország* (Gabriella Keczer, Erzsébet Sulyok); *Magyar Hírlap* (Erika Zádor); *Vasárnap* (Károly Nagy); *Reform* (Ella Bús); *Természetgyógyász Magazin* (Mária Markóczi); *Elixír* (Ildikó Molnár); *Képes Újság* (Miklós Gelléri); *Új Magyarország* (Attila Csarnai, János Orosz); *Új Demokrata* (Gábor Matúz, Károly Bánhidi); *Blikk* (Zsuzsa Bihari); *Fejér Megyei Hírlap* (Erzsébet Móré); *Pannon Napló* (Barbara Vági); *Hajdú-Bihari Napló* (Ádám Dombrovszky); *Kisalföld* (Csaba Gülch); MTV1; NapTV; A3; Vízöntő; TV2; Duna TV.

Last but not least, I would like to express my heartfelt thanks to my friends and colleagues: Zoltán Müller; Ildikó Budai; Kinga Kádár; István Havrancsik; Gábor Molnár; Judit Ludvai-Liptákné; László Lengyel; László Szalay; Gizella Budai; Robert Sobel; Ferenc Túri; Miklós Pereházy; András Dobák; Mihály Vadas; András Márton. Their generous work, advice and assistance have helped our project through many difficult stages.

"Could the cure to cancer be as simple as drinking water? This could be a possibility thanks to research done by biologist Gábor Somlyai..."
    The Northern Star /DeKalb, Illinois/

Deuterium depletion opens new perspectives in cancer treatment and prevention offering a completely safe and non-invasive treatment modality.

In Hungary all research and development, including drug registration and other product development, based on the deuterium depletion technology is exclusively managed, organised and financed by HYD Ltd. for Research and Development.

HYD Ltd. is accessible at:

Postal Address: 1539 Budapest
            POB 695.

Telephone: +36 1 309 7400
Telefax:+36 1 319 8976

e-mail: info@hyd.hu
home page: www.hyd.hu

Owing to HYD Ltd.'s product development, three kinds of deuterium depleted products are available, at present, in Hungary.

VETERA®-DDW-25 A.U.V

The world's first anti-cancer drug based on deuterium depletion. This efficient veterinary drug has been on the market in Hungary since 2000, which vets are regularly using to cure tumorous pets with. The D-concentration of Vetera® is 25 ppm.

PREVENTA®-105

Preventa®-105 is a deuterium depleted, carbonated drinking water. Marketed since 2000 as a healthy food product it is in fact used in therapy and prevention as a dietary supplement, in addition to regular regimens. (Anecdotal references are available.) The D-concentration of Preventa® is 105 ppm.

YUVAN

This cosmetic product line uses DDW as a key component. Dermatological tests prove its efficacy in revitalising and rejuvenating skin.

Printed in Hungary

Akadémiai Nyomda, Martonvásár